ECONOMIC BOTANY: AN INTRODUCTION FOR UNDERGRADUATES

ECONOMIC BOTANY: AN INTRODUCTION FOR UNDERGRADUATES

Prof N.C. Gautam

Delve PUBLISHING

www.delvepublishing.com

Economic Botany: An Introduction for Undergraduates

Prof N.C. Gautam

Delve Publishing
224 Shoreacres Road
Burlington, ON L7L 2H2
Canada
www.delvepublishing.com

Email: orders@arclereducation.com

ISBN: 978-1-77469-800-6 (Hardcover)

ABOUT THE AUTHOR

Prof. N.C. Gautam is M.Sc. Ph.D. in Botany-Genetics & Plant Breeding. He is a former Vice-Chancellor, M.G.C.G.V., State University Chitrakoot, Satna (M.P.) and VBS Purvanchal University, Jaunpur, UP. He has also worked as a Dean, Post graduate studies. NDUAT, Faizabad and Dean & Faculty of Hort., NDUAT, Faizabad. He has over 49 years experience in academics and more than 46 years of experience in teaching of Post Graduate and Under Graduate courses at University level. He also has 48 years experience, guided more than 8 Ph.D. students and 15 M.Sc Ag theses. He has published 20 books and many research papers. He is a Fellow of Indian society of Vegetable science, Fellow of Indian Botanical Society and Fellow of CHAI. He has received 15 awards and honors at national and international level. He has also worked on improving 30 Varieties of rice & vegetable crops besides other viable technology. He has also served as Chairperson/member of Expert Committee-UGC, Chairman/Member in Vice Chancellor Search Committees for many universities and Chairman-UGC and ICAR State Universities.

TABLE OF CONTENTS

LIST OF FIGURES

ABSTRACT

This book contains the most up-to-date data on the most important crops grown in various regions. Extensive discussion is given to the origins of agriculture, the evolution of agricultural plants, the discovery of plants, the migration of plants, the domestication of plants, and the proliferation of plants. Following a study of the ancestry and diversification of cultivated plants, we turn to a review of tropical, subtropical, and temperate crops used for food, beverages, spices, and medicines, as well as plant insecticides, timber plants, and essential oil-producing plants. Products from these crops range from food and drink to spices and even medicine. Plants that are used to make essential oils are also discussed. The beneficial impacts of a number of plants are examined in light of their unique genetic and evolutionary properties. Commercially important subtropical and temperate plants, as well as petro-crops, ethnobotanical studies, and biodiversity conservation, are discussed in this volume. The most important plant species belonging to each group are identified, and their significance is discussed at length.

This book is an excellent resource for students of botany, food science, nutrition, forestry, agriculture, horticulture, plant breeding, and environmental science. One of the book's strongest selling qualities is that it was written by a person who has devoted their entire life to the field of economic botany. Extensive photographs and line drawings are provided, all of which include descriptive captions.

The author did a lot of research in the field in Africa and studied plants that thrive in arid climates. The final output is a trustworthy and all-encompassing piece of literature that covers many different areas. The importance of proper nomenclature and categorization in the field of economic botany is highlighted due to the author's research and general interest in plant taxonomy. Economic botany in its broadest sense is within the purview of this discussion. Upon learning that ecophysiology, plant breeding, environmental preservation, and ecology will all play a role, the author became quite interested in the topic. This means that the book goes into greater depth than the vast majority of other works on the subject. The author was thrilled to see that the book covered the topic of using fungi and algae, both of which are now widely believed to no longer belong to the plant kingdom. Those interested in plant and fungal applications will find much to like in this book, as it covers a wide range of topics. Economic botany is becoming increasingly well-known for the role it plays in promoting ecological health and long-term prosperity. This is a great resource for anyone interested in these areas.

PREFACE

Economic botany is, by definition, the way people interact with plants. The field of economic botany is closely related to that of ethnobotany, which derives from two Greek roots: ethnos (or races: cultural group) and botanikos (of herbs), which can also refer to the study of a race's or people's knowledge of plants. This volume examines human-plant interactions from diverse perspectives, drawing on basic botany as well as archaeology, sociology, and ecology to explain these interactions and how they affect plants, society, and our ever-changing planet.

It focuses on plant cultivation's processes as well as its products. The authors ask about how groups learn about useful plants and share their knowledge. For instance, potatoes are staple food for many indigenous diets in the South American Andes. Researchers are curious about who first ate this vegetable and why they thought it might be tasty and nutritious despite the fact that the potato plant's leaves and stems are poisonous. What led these cultures to believe that beneath the surface might be something useful? How could they share their insight and with whom?

The use of plants is further examined in this volume. In the past, this meant making lists of cultures and the plants that they preferred to use for food, clothing, housing, medicine, ritual, or beauty. We typically use products from just 300 species in 20 of the approximately 250,000 plant species that are sub-divided into 460 families; only a minuscule part of what's accessible. A single plant can frequently serve multiple purposes. The coconut palm is a great illustration of the adaptability of plants. It is grown in the tropics, where it goes by a variety of names, such as pokok seribu guna, which means "tree of a thousand applications" in Malay. The plant's leaves, which are laced into thatch roofs and mats, the tasty fruit and sap, and the roots, which are processed to treat dysentery and bad breath, are all used. Economic botanists continue to catalog plant uses today, but they also seek to find new ones by searching for medicines like anti-cancer agents or testing with ways to enhance current cultivation to make it more sustainable or practical.

The processes of domestication and the connection between natural and human selection of particular plant traits are two important areas of study in the evolution of cultivated plants. Understanding how domestication may have altered a plant species over time requires knowledge of botany. In addition, ethnobotanists seek guidance on this process from fields like history, archaeology, and even linguistics. Take, for instance, maize. Physical evidence, such as DNA and stem and grain morphology that are similar, have demonstrated that maize is related to wild grasses in Central America and Mexico. This is in line with what archaeology and history have to say about how this plant was first grown there 7000 years ago. Additionally, the name suggests the crop's spread across cultures. Ma-hiz, an Arawak word, is translated into Spanish as "maize." This grain was first introduced to Europe by the Spanish in the Caribbean islands; beginning its eventual global spread. Economic botanists are also increasingly concerned about the effects of human activity on the geographical and biological diversity.

Chapter

1

BRANCHES OF ECONOMIC BOTANY

Contents

1.1: INTRODUCTION

Botany focuses on plants as a natural science. Botany, also called "plant science," is divided into three categories: core topics, which examine plant life's natural phenomena and processes, classification, and diversity; applied topics, which examine how plants can be used for economic gain in horticulture, agriculture, and forestry; and organismic topics, which examine plant groups like algae, mosses, or flowers (Kochhar, S. L. 2016).

Economic botany examines human-plant connections. Ethnobotany comes from the Greek words ethnos (race, people, or culture) and botanicals (of plants). Ethnobotany studies a race or people's plant lore. Economic botanists examine human-plant connections. Economic botany is as widespread and diverse as the world's people and plants. Economic botanists study people and plants from many aspects. Researchers use archeology, sociology, ecology, and basic botany to understand these links and their consequences on plants, society, and our dynamic planet. Economic botany examines plant agriculture's processes and products. Scientists wonder how plants learn and spread important knowledge.(Kochhar, Suraj Lal *2016).*

Plant-based applications: Often, one plant serves many roles. The coconut palm exemplifies botanical adaptability. Pokok seribu guna, Malay meaning "tree of a thousand uses," is one of its many names. The leaves are woven into thatch roofs and mats, the fruit and sap are delectable, and the roots are used to heal diarrhea and poor breath. Economic botanists continue to identify plant applications, but they seek to find new ones by exploring medicines like anti-cancer medications or improving agriculture.

Systematics, ecology, evolution: Methods of domestication and human and natural selection of plant features are studied in the development of domesticated plants. Botany is needed to understand how domestication has affected a plant species. Ethnobotanists consult history, archaeology, and language experts to understand this process.

Global trends: Many ethnobotanists worry about how human activities damage the landscape and biodiversity. Humans affect every ecosystem they inhabit. Landscape modification isn't limited to established agricultural nations. Environments that look untouched and abandoned may be human-managed. Native Americans ignited grassland and forest fires to cultivate and hunt. These tactics were used before European colonists cleared forests and farmed unclaimed land.

Human activities may have a detrimental influence on the environment by disrupting vital processes or overharvesting critical species. Overusing natural resources often leads to further decline and less variety. Due to overharvesting, European truffles are becoming difficult to locate. Truffles are growing more precious; therefore, hunters must work harder to discover rare ones. Even fewer fungi survive to grow next year's harvest.

Economic Botany's simple explanation hardly scrapes the surface of their achievements. Along with existing human populations and plant species, they also examine historical civilizations and yet-to-be-cultivated flora.

1.2 AGRONOMY

Agronomy involves producing plants for food, fuel, fiber, chemicals, recreation, or land conservation. Agronomy examines plant genetics, physiology, meteorology, and soil science. Biology, chemistry, economics, ecology, earth science, and genetics are applied. The subset of agronomists.

Agronomy studies crop genetics/physiology, crop management, and cultivation environments, especially soils. To produce crops in productive locations, their genetics and physiology must be understood. Ideal growing season duration, temperature range, and moisture needs. Certain crops, like soybeans, have day-duration demands that determine where they're grown. Many crops have different growing seasons, hence cultivars with designated seasons are made. To optimize the growing season, light-sensitive soybeans are planted from north to south. Latitude-dependent sunlight hours determine when flowers blossom.

Soybean seed must mature in the pod before flowering. Most crops are annuals, meaning they grow in a single season. Others are perennials grown as annuals. This includes sorghum. Agronomists study planting timing, row spacing, seeding rates, and crop nutrition. Cultural practices. Agronomy combines these aspects into a sustainable production system as technology develops. Crop rotation refers to the order and diversity of crops cultivated. All the information is a cropping system.

This agronomy issue involves selectively breeding plants to create optimal crops. Plant breeding has increased maize, soybean, and wheat yields and nutritional value. New plant species have also evolved. For example, Triticale was created by crossing rye and wheat. Triticale provides more protein than wheat and rye. Agronomy has helped fruit and vegetable studies. Plant breeding for turf grass development has reduced fertilizer and water inputs and produced disease-resistant turf varieties (Sparks, Donald

L 2012). Biotechnology helps agronomists develop desirable traits faster. Biotechnology requires field testing of newly created agricultural kinds. Agronomic biotechnology is being employed for non-food purposes. The oilseed is used largely for margarine and other food oils, although it may be used to make detergents, alternative fuels, and petrochemicals. Agronomists study sustainable ways to boost soil production. They examine soils for vital plant nutrients. Nitrogen, phosphorus, potassium, calcium, magnesium, and sulfur are investigated macronutrients. Micronutrients like zinc and boron are also evaluated. Regional labs measure organic matter, ph, and nutrition retention (cation exchange capacity). Agronomists will assess lab results and propose how to adjust soil nutrients for optimal plant growth. (Kochhar, S. L. 2016).

Figure 1.1: Agricultural scientist.

Source: US Navy 060427-N-1825C-001 Combat Systems Sciences and Technology Program Naval Postgraduate student Ensign Tom Dunbar, works with an autonomous robot originally designed to maneuver in agricultural settings. jpg: By U.S. Navy photo by Javier Chagoya - This image was released by the United States Navy with the ID 060427-N-1825C-001 (next). This tag does not indicate the copyright status of the attached work. Public Domain, https://commons.wikimedia.org/w/index.php?curid=8238742

Agronomists maintain soil and decrease wind and water erosion. Contour plowing reduces soil erosion and conserves precipitation. Agronomists want to use soil to tackle other problems. Human and animal feces disposal, water contamination, pesticide accumulation in the soil, and conserving the land for future generations by using paddock fires. No-till farming, soil-binding grasses on steep slopes, and contour drain up to 1 m deep are pasture management practices. Agro-ecology is the ecological management of agricultural systems. This problem relates to sustainable agriculture, organic farming, alternative food, and cropping systems.

Agronomy affects cabbage's antioxidant content. Soil type, fertilizers, and agricultural practices can affect crop antioxidant levels. Organic farming has gained popularity in recent years as farmers use natural organic fertilizers and management techniques to improve soil quality and plant nutritional content. Organic fertilizers and blends that contain organic and mineral fertilizers are replacing traditional chemical fertilizers, which affects cabbage's antioxidant content (Kochhar, S. L. 2016).

Organic and organic-mineral mix fertilizers have a higher antioxidant capacity than mineral or farmyard manure. Organic fertilizers result in a greater concentration of total phenolics in cabbage cultivars and a 40% rise in flavonoids compared to typical fertilizers such as urea, magnesium sulfate, potassium, and phosphorus. Organic fertilizers activate both the acetate and shikimate pathways, resulting in greater phenolic component synthesis.

Vermicomposting has recently caught farmers' attention as a soil conditioner to improve soil and crop quality, affecting antioxidant content and growth of numerous crops, including cabbage. Vermicomposting uses earthworms to turn organic waste into plant-friendly compost. Vermicomposting materials and earthworm population affect cabbage's antioxidant content. Three vermicomposting combinations were utilized to prepare the soil for cabbage plants: one with cow dung and vegetable waste, another with leaf litter and cow manure, and the third with all three. Vermicomposting and Pontoscolex corethrurus improved cabbage's vitamin C and sugar levels compared to inorganic treatment. The vermicompost mixture of cow manure and vegetable residue increased cabbage vitamin C by 12%, while the mixture of leaf litter and cow manure increased cabbage sugar by 57% compared to inorganic-treated soil.

Organic fertilization and vermicompost boost cabbage's antioxidant and nutritional characteristics. Vermicompost prepared from cow dung and earthworm *Eisenia fetida* increased soluble sugar and protein content of Chinese cabbages by 62% and 32%, respectively, as well as vitamin C (200%), flavonoids (17%), and phenols (25%). Chinese cabbage cultivated on soil treated with vermicompost and organic fertilizer had higher levels of leucine, aspartate, lysine, and alanine than those grown without vermicompost.

To maximize cabbage's phytochemical and nutritional profile and health benefits, it's important to consider the growing season, soil type, fertilizer choice, and agricultural techniques (Kochhar, S. L. 2016).

1.3 ETHNOBOTANY

Ethnobotany studies plants and their uses based on local culture and traditional knowledge. An ethnobotanist documents local behaviors, such as using native plants as medicines, food, intoxicants, and garments. Ethnobotany is the global study of plants.

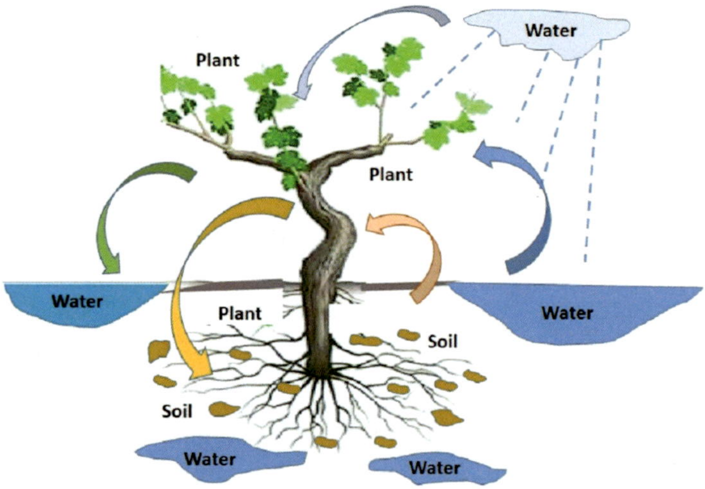

Figure 1.2: Shows the process of absorption, cohesion and transpiration of water in a tree.

Source: Gavrilescu M. Water, Soil, and Plants Interactions in a Threatened Environment. *Water*. 2021; 13(19):2746. https://doi.org/10.3390/w13192746

Ethnobotany has moved from gathering knowledge to applying it to modern society, mostly in the form of pharmaceuticals. Intellectual property rights and benefit-sharing are crucial in ethnobotany. In the 19th century, ethnobotany had more colonial ambitions than trade economics, documenting plants and their users. All of this research led to the creation of "aboriginal botany," the study of plants used by aboriginal peoples for food, medicine, textiles, and decoration. Modern ethnobotany stresses indigenous or traditional plant knowledge. Taxonomy, cultivation, and usage as food, medicine, and shelter are included(Kochhar, S. L. 2016). Early ethnobotanists studied how 'other' cultures used plants, although ethnobotany isn't necessarily about that. Not just indigenous cultures research medicinal herbs. Ethnobotany in plant selection requires detailed documentation of indigenous communities' relationships with plants. It's a challenging process that requires experts in anthropology, botany, ecology, pharmacology, languages, medicine, and ethnography. Ethnobotany examines human-plant relationships. Ford15 says ethnobotany has a common discourse but no unifying philosophy. Recognizing the reciprocal and dynamic human-plant relationship is crucial (Jain, S. K. 2010).

Ethnobotany combines botany and anthropology to study how indigenous people use plants. Combines plant science with ethnology. Because botanical knowledge is multidisciplinary, it's called a "multidisciplinary science." In recent years, ethnic groups and their expertise have been connected with ethnomedicine, ethnotaxonomy, ethnotoxicology, ethnoecology, ethnopharmacology, ethnopediatrics, and ethnoforestry.

Rudimentary medications created from traditional medicinal plants help many physical and mental diseases. Old writings, paintings, pottery, and plant fragments discovered in jars or midden heaps reveal prehistoric ethnobotany (trash dumps). This data can reveal people's agricultural and cultural advancement. Ethnobotanists sometimes reside in the community they are investigating for lengthy periods to observe all elements of their existence, including mythology, religious rites, and language, to uncover the exact plants used and the techniques involved in their production. Travel diaries, early botanists' field notes, and other writings give historical agricultural and medicinal expertise (Jain, S. K. (2010).

1.4 GENETIC RESOURCES

Genetic resources are genetic material with current or future value. Genetic material can be anything with functional units of heredity, whether it comes from a plant, animal, microbial, or other sources. Genetic resources are one of the three types of biodiversity. Any biological material that contains genes or metabolic material that may be produced from genes is referred to as a genetic resource. They are covered under the Nagoya Protocol when used for research or product development. Except for humans, all species' genetic resources within the respective geographic ranges of the contract parties are covered under the Nagoya Protocol. Genetic resources are not specifically mentioned in the Nagoya Protocol, although the Convention on Biological Diversity describes genetic resources and genetic material as follows: Genetic material is anything of plant, animal, microbial, or another origin that contains functional units of heredity. Genetic material that is valuable in the now or the future is referred to as a "genetic resource." Biological substances whose genes or the metabolic byproduct derived from them are the focus of research and/or product development are included below. Differing nations may, however, have different perspectives on genetic resources. If different nations' definitions have different meanings, the country of origin will be utilized.

Genetic resources from any biological creature, including whole genomes, genes, and even gene segments, that are currently or potentially valuable to humans are considered a provisioning ecosystem service. It is morally required to preserve them since they are essential to the evolution of ecosystems, their resilience, and persistence. In addition, they play a crucial role in domesticating or enhancing species that are used to produce food, textiles, medicines, and other goods for the benefit of humans. However, both natural and artificial influences, which can be slow (like climate change) or rapid (like genetic engineering), are endangering genetic resources (e.g., fire).

Crop plant genetic resources serve as an essential genetic resource, providing a reservoir of crucial genes for various survival attributes since they have evolved over thousands of years of struggle and survival against the environment. The wild and weedy relatives of agricultural plants live in harsh environments, thus they exhibit several adaptation-related traits in addition to resistance to a wide range of significant biotic and abiotic stresses. They must thus be collected, characterized, conserved, and used in agricultural improvement initiatives.

Genetic resource loss is mostly caused by overexploitation, land use change, alien species invasion, pollution, and fragmentation. There is potential for ex situ (outside of natural ecosystems) conservation of genetic resources (in artificial environments). Although pollen banks, *in vitro* preservation, cryopreservation, DNA storage, microbe culture collections, and field gene banks are also employed, seed banks are the most popular ex-situ conservation technique. Genetic markers and common garden experiments are the best tools for describing and assessing genetic resources (Kochhar, S. L. 2016).

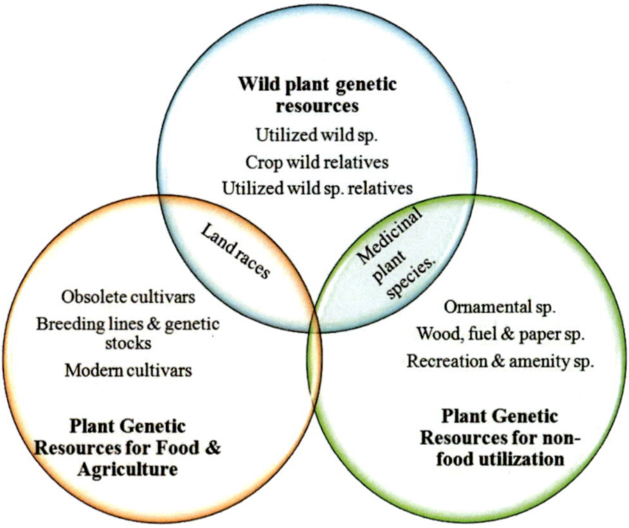

Figure 1.3: Genetic resources cycles.

Source: Priyanka V, Kumar R, Dhaliwal I, Kaushik P. Germplasm Conservation: Instrumental in Agricultural Biodiversity—A Review. *Sustainability*. 2021; 13(12):6743. https://doi.org/10.3390/su13126743

1.5 HORTICULTURE

Horticulture encompasses cultivating plants for food, medicine, comfort, and ornament. Horticulturists grow trees, lawns, flowers, fruits, nuts, vegetables, and herbs. Horticulture helped nomadic peoples become permanent or semi-sedentary. Horticulture is separated into divisions that focus on various plants and meals. Multiple organizations teach, develop, and promote horticulture to preserve it.

Plant propagation and care promote plant growth, yields, quality, nutritional value, and resistance to insects, diseases, and environmental challenges. Included are plant conservation, landscape restoration, soil management, garden design, building and care, and arboriculture. Horticulture stems from the Latin words for "garden" and "cultivation," Hortus, and culture. Horticulture excludes intense crop cultivation, large-scale agricultural output, and animal husbandry. Horticulture emphasizes small plots with various crops, whereas agriculture concentrates on one core crop at a time.

Plant types

Horticulture has many subfields. They include:

- Vegetable production (olericulture)
- Pomology is the cultivation of fruits and nuts.
- Grape cultivation (primarily meant for winemaking)
- Floriculture: cultivating lovely flowers.
- Turf management: Growing and maintaining grass for sports, enjoyment, and beauty.
- Arboriculture: the cultivation and maintenance of trees, shrubs, vines, and other perennial woody plants for landscape and amenity.
- Landscape horticulture encompasses plant selection, cultivation, and maintenance.
- Postharvest physiology: handling harvested horticultural crops to reduce deterioration during storage or transit.

Horticulture stems from the Latin terms for "garden" and "cultivation," Hortus, and culture. Horticulture doesn't involve large-scale farming or animal husbandry. Horticulture doesn't involve large-scale farming or animal husbandry. Agriculture focuses on a single huge primary crop, but horticulture utilizes small plots with a range of crops. Horticulture has been studied and practiced for thousands of years, and it may have contributed to the transformation from nomadic to sedentary horticultural cultures. Each form of horticulture focuses on producing and preparing various plants and meals. Several organizations teach, encourage, and support horticulture to maintain science (Adams, C. R. 2012).

Plant propagation and nurturing boost plant growth, yields, quality, nutritional value, and tolerance to insects, diseases, and environmental challenges. Horticulture yields more than field crops per hectare. Horticulture crops are renowned for their nutrients. Fruits and vegetables offer vitamins and minerals. Horticulture improves the environment's appearance. Horticultural crops benefit small farmers. Horticulture offers diverse crop variations. Horticultural crops decrease waste, preserve soil and water, and enhance farmers' incomes. It blends scientific, technological, and manufacturing procedures to fulfill clients. Horticulture methods vary. Let's review some major gardening branches.

Arboriculture researchers cultivate and manage trees, shrubs, vines, and other perennial woody plants. It investigates how plants develop and adapt to their surroundings and culture. Arboriculture entails choosing, planting, preparing, fertilizing, managing insects and diseases, pruning, shaping, and removing. Arborists look for individual woody plants and trees in gardens, parks, and other populated places for the delight, safety, and benefit of humans.

Landscaping

Turf management, or pitch care, maintains a playing surface. Turf management entails developing and maintaining turf. Sports, entertainment, and amenity turf grass upkeep. Floriculture is the cultivation of blooming plants for gardens and floristry. Floriculturists develop new varieties. Floriculture crops include bedding plants, houseplants, floral gardens, container plants, and cut flowers. Floriculture plants are herbaceous, unlike nursery plants. Landscape horticulture generates outdoor settings, monuments, and constructions with environmental, social-behavioral, or aesthetic aims (Kochhar, S. L. 2016).

It combines the systematic design and general engineering of varied structures for construction and human use, as well as an investigation into existing social, ecological, and soil conditions and processes in the landscape, and the creation of extra measures to achieve the intended aims.

Olericulture

Olericulture grows non-woody plants for human consumption. The science of cultivating, preserving, refining, and selling vegetables is referred to as olericulture. It includes cultivar selection, seedbed preparation, and seed and transplanted vegetable crop establishment. It also includes industrial and non-traditional vegetable crop production, such as organic gardening and organic farming, sustainable agriculture and horticulture, hydroponics, and biotechnology, as well as organic gardening and organic farming, sustainable agriculture and horticulture, hydroponics, and biotechnology. Pomology, a discipline of botany, analyzes fruit and its cultivation. The primary themes of pomological research are fruit tree growth, enhancement, culture, and physiological tests. Fruit tree enhancement attempts to improve fruit quality while simultaneously cutting harvest times and production costs.

Viticulture is a horticulture science that deals with the production and harvesting of grapes. The viticulture technique encompasses monitoring and managing pests and diseases, fertilizing, irrigation, canopy management, fruit production and characteristics, selecting when to harvest, and winter grape cutting. Oenology is a field of horticulture that investigates wine and winemaking.

Post-harvest physiology is the scientific study of the physiology of living plant tissues after harvest to conserve plant quality and reduce deterioration. It offers immediate benefits in post-harvest handlings, such as selecting the appropriate storage and transportation conditions to maximize shelf life.

1.6 MICROBIOLOGY

Microorganisms are vital to nearly every activity on Earth. Microbes are crucial because they exist within, on, and around humans. The study of microorganisms is known as microbiology. Microorganisms include bacteria, archaea, viruses, fungi, prions, protozoa, and algae. These bacteria are crucial for the nitrogen cycle, biodegradation and biodeterioration, climate change, food spoilage, disease etiology and control, and biotechnology.

In addition to food and drink, bacteria may make biofuels, life-saving pharmaceuticals, and other products. Microbiological research is crucial for solving several global goals and barriers, such as preserving food, water, and energy security for healthy people on a habitable globe. Questions like "How diverse is life on Earth?" and "Does life exist elsewhere in the universe?" can be addressed with the assistance of microbiological study. The microscope was a major influence on microbiology. Either life is biotic or spontaneous (Adams, C. R. 2012).

Microbes, according to proponents, were independent and did originate in this fashion. In the early 1800s, germs may develop new ones (biogenesis)

Microbes

The major kinds of microorganisms include viruses, bacteria, fungi, algae, and archaea. Bacteria include archaea and eubacteria. Microbiology was established from the study of microorganisms. The change from bacteriology to microbiology evolved as a consequence of these methodologies, which were first employed to research bacteria but have subsequently been improved for use with all organisms (Kochhar, S. L. 2016). All bacteria are prokaryotic, which means that their nuclei are not membrane-bound. The genetic material in each cell, or DNA, is a long, folded thread with no set place. It was believed that all bacteria evolved in the same way up until the late 1970s. Two bacterial groups shared an ancestor, according to ribosomal RNA from different bacteria. The three primary groupings are—the eukarya (the eukaryotic cells), the archaea (the bacteria that diverged from other bacteria at an early stage of life), and the eubacteria. Today, eubacteria are acknowledged as actual bacteria and belong to the realm of bacteria. Intriguing commonalities between these three groups were revealed when the DNA sequences of various bacteria were studied. As a result, it is tough to trace the roots of present ailments. Even other microbes have displayed features specific to certain taxonomic groupings. The "missing link" in the global nitrogen cycle, an anaerobic ammonia-oxidizer, was located in 1999. This Planctomycetales bacterium contains eukaryotic-like internal structures, an archaean cell wall, and yeast-like budding.

Algae

The DNA of eukaryotic bacteria is enclosed in a nuclear membrane, much like in plant and animal cells. Among eukaryotic animals include fungi, protozoa, and algae. Algae, protozoa, and certain lower fungi are examples of protists (order Protista, often labeled Protoctista)

Fungi

Like algae, fungi are eukaryotic animals with strong cell walls. Some are microscopic, while others develop bracket fungi and mushrooms on moist wood. Unlike algae, the fungus cannot create photosynthesis because they lack chlorophyll. Instead of eating, the fungus takes up nutrients from its surroundings. Yeasts are unicellular bacteria, whereas molds are multicellular germs that form minute filaments. Cylindrical mold cells can generate threadlike filaments (hyphae) that may include spores. They are microscopic fungi called hyphae. A visible mycelium emerges when several hyphae cluster together, as on bread or fruit jelly. Unicellular yeasts include spherical, egg-shaped, and filamentous yeasts. Yeasts ferment carbohydrates in wine and bread, generating alcohol and CO2.

Protozoa

Protozoa are single-celled eukaryotic microorganisms. Other protozoa have expanded in length. Others modify form during their lives. Cells range in size from 1 m to 2 mm (visible without magnification). Protozoa, like animal cells, lack cell walls and may move and absorb food; nonetheless, many phytoflagellate protozoa are plantlike and produce energy through photosynthesis. Animal-like traits are seen in protozoan cells. Some swimmers deploy hair-like cilia or flagella to propel themselves. Images captured with a microscope reveal their darting action in pond water droplets.

1.7 NUTRITION

Nutrition is the process by which living organisms ingest food to grow, maintain, and reproduce.

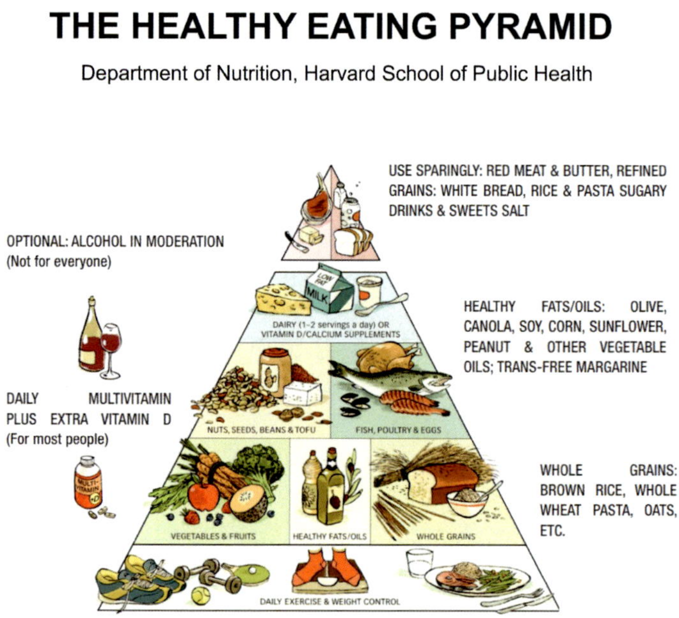

Figure 1.4: A generalized healthy diet and lifestyle pyramid.

Source: Harvard Dietary Pyramid [7]. Copyright © 2008. For more information about The Healthy Eating Pyramid, please see The Nutrition Source, Department of Nutrition, Harvard T.H. Chan School of Public Health, www.thenutritionsource.org (accessed on 1 February 2021), and Eat, Drink, and Be Healthy, by Walter C. Willett, M.D. and Patrick J. Skerrett (2005), Free Press/Simon & Schuster Inc.: Fernandez ML, Raheem D, Ramos F, Carrascosa C, Saraiva A, Raposo A. Highlights of Current Dietary Guidelines in Five Continents. International Journal of Environmental Research and Public Health. 2021; 18(6):2814. https://doi.org/10.3390/ijerph18062814

Food is consumed by most organisms for numerous purposes. It supplies ingredients that are digested to create energy for nutritional absorption and translocation, cell material synthesis, movement and motility, and waste product excretion, among other tasks. All structural and catalytic cell components are given by food. The nutritional ingredients required by living things vary, as do the means via which they are created or obtained, and the role these chemicals play in their cells. The nutritional process and nutrients essential to maintaining life across the living world follow broad patterns. Patterns are explored in this article.

Biological feeding

The taxonomy of living organisms is established by how they digest food. Green plants and some bacteria, for example, require simply inorganic elements for development, whereas heterotrophic organisms require both inorganic and organic components. Nutritional patterns are another category. One classification strategy categorizes animals depending on their energy source. Phototrophic creatures, also known as photosynthetic organisms, turn light energy into chemical energy, whereas chemoautotrophic species, also known as chemosynthetic organisms, utilize inorganic or organic chemicals. The organism is lithotrophic if its electron-donor components are inorganic or it is organotrophic if they are organic (Insel, P. M. 2014). These patterns are also used to define organisms. Light energy, for example, is used by higher plants, with water serving as the final electron source. Photosynthetic bacteria that cannot use water as an electron source are categorized as photoorganotrophs. Animals are chemoorganotrophs, meaning they obtain their energy from chemical compounds and their electrons from organic substances.

Nutrition from vegetables

Plants do not require organic components for food, although they do constitute the bulk of their tissues. Photosynthetic systems gather solar energy and use it to create nutrients from CO2 and water. Phosphorus (as phosphate), chlorine (as chloride ion), potassium, sulfur, calcium, magnesium, iron, manganese, boron, copper, and zinc are all obtained by plants from the soil around their roots. Plants require nitrogen ions (NO3 or NH4+). Inorganic compounds such as iodides, cobalt salts, and selenium salts are also absorbed.

Rain and low temperatures disintegrate rocks on the Earth's surface, releasing nutrients into the soil. Rocks are mostly alumina and silica, but they also comprise minerals that plants require. Plants, animals and their waste decompose and supply nutrients to the soil. A spadeful of soil may appear inactive, however, it contains millions of bacteria that breakdown organic compounds and liberate mineral salts. Two types of bacteria are capable of converting ambient nitrogen into nitrate ions. Rhizobium bacteria can be found in the roots of leguminous plants such as peas and beans. In the flooded paddy fields of Southeast Asia, cyanobacteria (blue-green algae) fix nitrogen and contribute to the development of rice.

Fertilizers are essential in intensive farming, where crops are harvested once a year and no animals graze. Traditional fertilizer was animal feces, also known as muck, which was made from soaking straw bedding and waste. Farmers have employed artificial fertilizers such as chalk (for calcium), rock phosphates, and guano since the 1800s. Guano for sale is a nitrate-rich mound of bird droppings. Fertilizers are comprised of nitrogen, potassium, and phosphorus. The bulk of nitrogenous fertilizers is made by combining nitrogen and hydrogen under high pressure with catalysts to generate ammonia (NH3). This may be injected into the soil as a gas or processed into solid fertilizers such as ammonium salts, urea, and nitrates.

Nutrition of Bacteria

These minute critters, which are generally seen as infection sources, play a key function in plant and animal life cycles. They, like bigger species, must digest their food, and their cell walls prevent huge molecules from entering. When bacteria are bathed in a sugary liquid, the sugars diffuse through the bacterial wall, forming larger molecules that assist inward diffusion. Bacteria must generate digestive enzymes to utilize larger molecules such as carbohydrates and protein (i.e., catalysts). Because many released enzymes and digested products move out from the bacterial cell, this is a cost function for a single organism. The method is less economical for hundreds of millions of related bacteria.

Bacterial dietary demands vary. Sugars and inorganic nutrients are required by some, such as plants. Some require oxygen to convert carbohydrates to carbon dioxide and water to harvest energy. Others are anaerobic (oxygen damages them in specific settings) and require sugar to ferment into lactic acid or ethanol and carbon dioxide, providing less energy but enough to fit their demands. After multiple generations in a nutrient-rich environment, certain bacteria have lost the potential to manufacture various essential compounds. Lactobacilli, which are present in raw milk, require all water-soluble vitamins and amino acids. As a result, they are employed to determine the nutritional worth of food.

Diet of animals

Plants are devoured by animals, according to observation. Even carnivorous creatures, such as lions, consume grazing animals and are thus indirectly dependent on vegetation.

- Herbivores

Animal digestive enzymes are incapable of digesting cellulose, the primary component of plant cell walls. Plant cell nutrients are not entirely absorbed. Many herbivores have acquired a pouch at the front of the stomach called the rumen to ferment eaten leaves as an evolutionary reaction. Fermented cud is regurgitated from the rumen by ruminant animals such as cattle and sheep so that the animal can chew it and disseminate ruminal fluid throughout the ingested content. Ruminal microorganisms convert cellulose into acetic acid and other short-chain fatty acids that can be turned into energy (Kochhar, S. L. 2016).

Protein is also secreted and destroyed in leaf cells; some are resynthesized as microbial protein in the stomach and small intestine. Ruminal bacteria also create water-soluble vitamins, which the host animal does not need to consume. Ruminal fermentation hydrogenates fatty food because the rumen is anaerobic. Excess hydrogen can be added to unsaturated fatty acids if it cannot be coupled with oxygen to generate water. Saturated fatty acids generate firmer fat deposits following absorption. At room temperature, beef fat (suet) is firmer than pig or chicken fat. Glycerol esters include short-chain fatty acids that make butterfat supple at normal temperatures. Ruminant fats lack necessary polyunsaturated fatty acids, making them undesirable as a dietary source.

Other herbivores benefit from hindgut fermentation because it assists in the digestion of green meals. Animals are the principal hosts of small intestinal enzymes and bloodstream absorption. To preserve water during droughts, the large intestine absorbs the majority of the remaining water. Before entering the big intestine, undigested food remnants ferment in the cecum, a tiny intestinal side pocket. Cecal short-chain fatty acids are absorbed and utilized by the large intestine. Horses, zebras, elephants, rhinoceroses, koalas, and bunnies are all suited here.

Ruminants digest high-fiber meals more effectively than hindgut fermenters. Because only indigestible wastes ferment in the cecum, hindgut fermenters do not lose energy like rumen fermenters. The tiny cecum permits these animals to be nimbler and dodge carnivorous predators.

- Carnivores

Carnivores form a tiny part of the animal kingdom since each species must devour a huge number of similar-sized animals to survive. Carnivores have fangs and claws that allow them to kill and tear apart prey, as well as digestive enzymes that break down muscle protein into amino acids that can pass through the small intestine. Carnivores do not require a fermentable gut.

Carnivores make utilization of animal fat. Calcium may be gained through chewing and swallowing tiny prey. Some carnivores, such as cats (family Felidae), are obligatory carnivores, which implies they cannot receive adequate nutrition from plants and microbes. Obligate carnivores lack the enzyme that transforms carotene from plants into vitamin A. Vitamin A is taken from the liver of their victim. Some required long-chain, highly unsaturated fatty acids cannot be produced by obligate carnivores from plant fatty acids.

- Omnivores

Omnivores are a broad collection of organisms with dentition and digestive systems suited to a concentrated diet. They lack a big sac or chamber for fibrous material fermentation. They can chew and digest meat, but they don't require it until there is no alternative supply of B12 (cobalamin). Humans, dogs, rodents, and the majority of primates are welcome. Omnivores have active bacterial flora in their small and large intestines, allowing them to absorb short-chain fatty acids but not vitamins. Some animals receive vitamins through coprophagy, which includes ingesting feces containing vitamins created by bacteria. Chickens are omnivores. They swallow without chewing, but the food is channeled to the gizzard, where seeds and other foods are crushed with swallowed stones.

Certain organisms are unable to synthesize organic nutrients and must consequently obtain them performed. There are carbs, proteins, and lipids. Vitamins are needed in low concentrations for catalytic or regulatory actions in metabolism.

Carbohydrates: Plant-made carbohydrates, which give the majority of the energy of the animal kingdom are the most significant nutrients in terms of quantity. Birds and small animals are lured to the sweet ripe fruit. The fruit's seed covers survive the animals' fast digestion, allowing the plant's viable seeds to multiply. Sucrose accumulates as an energy reserve in sugarcane stems and sugar beet roots; both are exploited to create table sugar.

Disaccharides have two glucose units whereas monosaccharides have one. Complex carbohydrates must be broken down into simple sugars that can be consumed and absorbed fast. Sucrose must hydrolyze even freely soluble sucrose to glucose and fructose. Piglets born missing this enzyme are unable to consume sucrose. Lactose is immediately digested in newborn animals, but most species, including humans, stop secreting lactase after weaning. Lactose is found solely in milk, which animals will not swallow after nursing.

Starch is a polysaccharide that is formed by the condensation of numerous glucose units, usually through connections that are broken down by digestive enzymes in bacteria and higher animals. Cross-linkages between basic chains in plant starches vary, resulting in more compact, resistant molecules. Cooking allows starch granules to expand, making them more digestible. Surprisingly, even carnivorous cats can benefit from finely crushed starch. Dry cat food includes up to 20% starch.

Plant cell walls are made up of cellulose. Cellulose is created from condensed glucose units, but the chains can lay flat due to a special sort of connection. These connections cannot be digested by vertebrates because they lack enzymes (Wickens, G. E. 2012). Herbivorous animals have digestive systems that allow bacterial fermentation of cellulose in the foregut (rumen) or hindgut, allowing them to benefit from cellulose metabolites, notably short-chain fatty acids. When pectin and hemicelluloses are hydrolyzed, they generate xylose and arabinose. These sugars are fermented by bacteria, but not digested by mammalian enzymes. Lignin, an indigestible phenolic polymer, is present in stiff plant structures. These compounds are dietary fiber.

Fats and oils (lipids): Some plants store energy in the form of fat, or oil, in their seeds. Animals store energy in the form of fat. Fats contain a larger concentration of energy than carbohydrates, with 9 and 4 kcal/g, respectively. A fat is made up of three fatty acids connected by a glycerol backbone. The physical features of fats are dictated by fatty acids. Liquid fats are present in biological tissues. The fats of warm-blooded mammals have a larger freezing point than fish fats. Plants that are frost resistant have a low freezing point. Fat is accumulated by organisms with little or no excess liquid; its freezing point is close to the organism's optimum viability. Fatty acids are characterized by chain length and saturation. The bulk of fatty acids comprise chains with 16 or 18 carbons. Cow's butterfat has a low freezing point due to its 4-carbon short-chain fatty acid butyric acid; the longer the saturated chains, the higher the freezing point of the acids and fats that contain them. Chain liquidity is boosted by unsaturated (double) bonds. Polyunsaturation lowers the solidity of lipids at room temperature.

For energy, animals may store or oxidize ingested fatty acids. Certain fatty acids are required for the production of phospholipids, which are found in cell membranes and nerve fibers, as well as certain hormones. Animals may manufacture fat from excess carbs, but not required polyunsaturated fatty acids such as linoleic and linolenic acid. Fatty acids are a crucial dietary component, not merely an energy source. Linolenic acid is present in the majority of the main vegetable oils. Cats lack an enzyme that transforms linoleic acid to arachidonic acid for the generation of prostaglandins and other hormones. Because arachidonic acid is not contained in plants, cats must ingest animal meat to thrive and breed.

Proteins: Proteins, which are large molecules with chains of 20 amino acids, are the fundamental organic material found in plant and animal tissue. Animals degrade protein into amino acids before it enters the blood. If plants have nitrate or other simple nitrogenous chemicals and sulfur for cysteine and methionine synthesis, they can generate their amino acids, which are essential for protein production. Some amino acids can be produced by animals from ammonium ions and carbohydrate metabolites, whereas others are needed. Methionine and phenylalanine are used to produce cysteine and tyrosine. Bacteria in the rumens of ruminant animals may synthesize all of the amino acids in a protein, ensuring that the ruminant's stomach continues to acquire high-quality microbial protein for digestion. Protein is necessary for growth. This is mirrored in the protein amount of the milk secreted during suckling. A piglet's birth weight doubles in 18 days and a sow's milk protein offers 25% of the total calories. Humans quadruple their birth weight in 180 days, although breast milk comprises just 8% protein. Animals fed diets lacking one key amino acid quit growing quickly (Wickens, G. E. 2012).

Adults need more protein than is lost through urine, feces, hair, and skin. Animal tissues are continuously "cycling over" their proteins—hydrolyzing and re-synthesizing them—but this doesn't explain the increasing protein consumption, since the amino acids released may be reused. When the body is short on protein, the enzymes that digest extra amino acids don't inactivate totally, but instead "idle." Adult animal diets, including human meals, provide ample protein to counteract idle losses. Evolution may have adapted idle rates to protein intake. So, adult rats eating a range of foods—some low in protein—need no more than 5% protein. Cats, whose old carnivorous diet was richer in protein, require 20% to counterbalance losses.

Vitamins: Vitamins are chemical compounds that serve a catalytic role in the body (usually as coenzyme or enzyme components) and must be acquired from the diet. Vitamin inadequacy can induce sickness in each organism. In young animals, vitamin shortages promote growth failure, different symptoms, and mortality. Vitamins are typically regarded as organic compounds that animals and humans must get from their food in tiny amounts. Vitamin A isn't present in plants, but carotene is, and most animals can convert it into vitamin A. Cats and other carnivores must eat animal tissues to acquire preformed vitamin. Niacin is not a necessity as most animals (cats being an exception) can generate it from tryptophan if the latter is present in excess.

1.8 PHARMACOLOGY

The scientific study of drugs and their effects on living organisms is known as pharmacology. Pharmacodynamics is the scientific study of how drugs function within the body. Pharmacokinetics is the study of drug metabolism and elimination from the body

Due to their role in drug discovery and development, medieval apothecaries were the precursors of contemporary pharmacologists. In the first half of the nineteenth century, pharmacology emerged as a separate profession from apothecaries, who merely acted as intermediaries between patients and drug manufacturers. After chemistry and biology made significant advances in the late 18th century, true scientific pharmacology developed. In the early nineteenth century, French and German chemists isolated morphine, strychnine, atropine, and quinine from plants. Antibiotics such as penicillin and hormones such as insulin and cortisol are only a few of the numerous new drugs that have resulted from pharmacological research since the end of World War II. Through chemical synthesis, pharmaceutical science is enhancing these and other medications. In pharmacology's pursuit of more effective drug administration, research involving large patient populations is essential.

In the early 20th century, pharmacologists discovered that the chemical structure of a compound determines its physiological effects. Since then, pharmacologists have been interested in the performance effects of modifying the molecular structure of a drug. Due to the prevalence of organic compounds in medicine, pharmacologists must be educated in organic chemistry.

Pharmaceutical and chemical companies conduct fundamental pharmacological research. Following the year 1930, pharmacological research flourished in the United States and Europe. Before promising new drugs to enter clinical use, industry pharmacologists test them. Before any conclusions can be drawn about a drug's effects on patients or its toxicity to humans, its effects on the systems and organs of laboratory animals must be studied. The pharmacologist does not conduct human testing to determine a drug's clinical efficacy; instead, doctors do so, but only after the drug has undergone extensive animal testing. Constant testing is necessary for regulating and standardizing the efficacy and purity of pharmaceuticals (Wickens, G. E. 2012).

Numerous pharmacology subspecialties

- Pharmacodynamics: The study of the long-term physiological and biological effects of a drug's concentration is known as pharmacodynamics. Understanding how drugs reach the brain is the focus of this field of study. Many pharmaceuticals produce unwanted side effects in addition to their intended therapeutic effect. This can be affected by the dosage of the medication. If you consume an excessive amount of magnesium, you may experience stomach cramps and diarrhea.
- Pharmacokinetics: The medical field of pharmacokinetics examines how drugs are absorbed, metabolized, and eliminated from the body. Medicines can be administered orally, intravenously, or intragastrically (into the bloodstream, through an IV). The lungs and sweat glands play significant roles in the elimination of drugs.

The following branches of pharmacology also utilize pharmacokinetics and pharmacodynamics:

- Clinical pharmacology is the study of how medications are used in the treatment and how factors such as age, pregnancy, disease, and drug combinations influence drug efficacy. Bioavailability refers to the proportion of a drug dose that is absorbed by the body.
- Toxicology is primarily concerned with the study of adverse drug

reactions. The program investigates the effects of therapeutic drugs and chemicals on the home and works lives of people.

Pharm students concentrate on pharmacology in their studies. Pharmacists have a vast understanding of pharmaceuticals. They assist physicians and nurses with patient care by administering medication and providing instructions and advice. To learn more about drug mechanisms and aid in drug design, pharmacologists conduct laboratory research on drugs. They research medications, whereas pharmacists provide the public with finished products and the necessary information (Wickens, G. E. 2012).

Chapter

2

THE RELATIONSHIP BETWEEN PLANTS AND PEOPLE: PAST, PRESENT, AND POTENTIAL USES OF PLANTS

Contents

2.1 INTRODUCTION

Plants not only provide food for humans, but they also play essential roles in our religious and spiritual practices and as ingredients in traditional medicines. Native forest dwellers collect plant materials for ritual or medicinal purposes, and the delicate curve of a tendril has inspired works of art. Plants play an essential role in human societies for reasons other than sustenance. Art from various eras emphasizes the importance of plants to humans. Ancient Pueblo people carved petroglyphs into the cliffs to demonstrate the significance of the maize plant (*Zea mays*) to their culture. Papyrus (*Cyperus papyrus*) is frequently depicted in Minoan art (2600-1100 BC), whereas lychees (*Litchi chinensis*) are frequently depicted in exquisite Chinese art. Nature has always served as a source of inspiration for the human race.

The types of relations include:

2.2 COEVOLUTION

There are deep evolutionary connections between humans and plants. The term "coevolution" was first used by Peter Raven, a leading expert in plant biology whose own work is featured in this issue (Raven, 2019), and his colleague Paul R. Ehrlich. Many plant-human interactions are included in this definition. The Australopithecus africanus, who lived 3 million years ago, is thought to have been very similar to modern humans. Australopithecus afarensis, a primate that lived in forests, ate tough nuts and underground storage organs of plants. The jaw of the Australopithecus was massive, and its skull was heavily ridged so that it could attach to the animal's massive muscles. Australopithecus could digest tough plant matter because of the shape of its skull, which developed in response to its diet. In order to better process plant foods, Australopithecus had a very long digestive tract.

Homo erectus, a human ancestor dating back 1.5 million years, differed from Australopithecus in several ways, including having a lighter skull, smaller teeth with thinner enamel, and a shorter digestive tract. The grasses and seeds that *H. erectus* would have hunted on the savannah are more nutritious and accessible to digest than the plants Australopithecus ate. Australopithecus's muscular or digestive systems were not as complex as those of *H. erectus*. The biochemical composition of the food they ate also shifted, reflecting these morphological changes in hominins. One of the most crucial biochemical pathways in plants is the shikimic acid (SA) pathway, which allows plants to biosynthesize three of the nine essential amino acids that humans cannot produce. Evolutionary history has taught us that in order to get the SA pathway derivatives we need; we must rely on plant foods. Pharmaceutically beneficial flavonoids and alkaloids are generated via the SA pathway as well. (Wardle, D. A. 2013). For instance, star anise (*Illicium verum*) and sweetgum (*Liquidambar styraciflua*) are the sources of SA, which is then used to create Tamiflu, a drug used to treat severe influenza symptoms. These biochemically-based dietary and therapeutic uses highlight the evolutionary closeness between plants and humans.

Humans also have more diverse taste buds than any other animal. This is further evidence of the crucial link between plants and humans, though this may be cultural rather than evolutionary. Grain crops like rice (*Oryza sativa*), wheat (*Triticum aestivum*), maize (*Zea mays*), or teff (*Eragrostis tef*), and legume crops like peanuts (*Arachis hypogaea*), broad beans (*Vicia faba*), and chickpeas (*Cicer arietinum*) are staples in the diets of many different cultures. In order to function correctly, the human body requires a particular set of amino acids called essential amino acids, which cannot be biosynthesized and must instead be obtained through the diet in the form of complete proteins. Cereals have a very balanced amino acid profile, while legumes are deficient in some but abundant in others. Getting all of the essential amino acids from just two plant sources is a remarkable dietary trait seen across many human cultures. Ascorbate is a fascinating example of this phenomenon (vitamin C). Humans cannot produce vitamin C, so traditional societies have to rely on food sources like fruit and vegetables to prevent a lack of this nutrient that can cause scurvy.

2.3 DOMESTICATION

The domestication of plants was a watershed moment in human history. Until about 15,000 to 20,000 years ago, hunter-gatherers ate only wild plants. These previously wild species became domesticated due to human contact, resulting in genetic changes. They looked very different from their wild ancestors when they were domesticated. Despite being related, Oryza rufipogon is rice's wild ancestor. O.'s long awns and prominent anthers are absent from the modern crop. Sativa was grown for seed dispersal. Most rice cultivars fertilize the seeds before flowering, which is thought to be the cause of these striking visual changes. Domestication has altered the genetics of many species to the point where they can no longer thrive or compete in the wild. As a result, they have become overly reliant on others. (Wardle, D. A. 2013).

Many species have been domesticated all over the world. Domestication occurs in several regions, including Asia, where rice and soybeans are examples of domesticated crops (*Glycine max*). Many cultivated crops grow in the Fertile Crescent, including African yams (*Dioscorea* spp.), cereal grains like wheat and barley (*Hordeum vulgare*), and starchy vegetables like corn and potatoes. The Americas are also known for growing maize and potatoes (*Solanum tuberosum*). Natives of eastern North America domesticated many species. According to Jared Diamond, Domestication has been the most significant technological advancement in the last 15,000 years (1997). The ability of our forefathers to domesticate and cultivate native plants altered how they used their surroundings. As a result of the increased availability of food, permanent settlements were established, populations increased, and divisions of labor were established. Because not everyone participated in traditional agricultural and hunting activities, a cultural change occurred. Teosinte (*Zea mays* ssp. parviglumis), the wild ancestor of maize (*Zea mays* ssp. Teosinte, corn, and *O. rufipogon* all have very distinct reproductive anatomy. Domesticated maize has only a few large leaves, two ears, and a single stalk, unlike the sizeable perennial teosinte, which has many flowers and flowering branches.

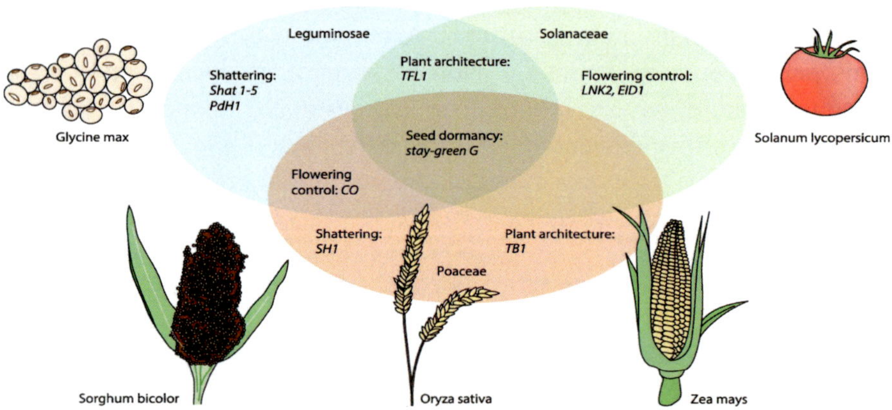

Figure 2.1: Examples of parallel selection of protein-coding genes across plant families. Depicted are: maize, *Z. mays*; rice, *O. sativa*; sorghum, *S. bicolor*; soybean, *G. max*; and tomato, *S. lycopersicum.*

Source: Rendón-Anaya, M., Herrera-Estrella, A. The advantage of parallel selection of domestication genes to accelerate crop improvement. Genome Biol 19, 147 (2018). https://doi.org/10.1186/s13059-018-1537-7

The term "artificial selection" was coined by Darwin to describe the process by which humans shaped the evolution of domesticated crops at the expense of their wild ancestors. Some plants were discovered to be superior by the first farmers in terms of taste, yield, seed retention, ease of cultivation, and robustness. People collected seeds from wild plants and grew them in gardens near their homes. (Wardle, D. A. (2013).

2.3.1 The Origins of Domestication

Many characteristics are linked to crop domestication. This is a common problem in domesticated plants when the seeds are edible but cannot be dispersed because the plant will be harvested before the next growing season. On the other hand, farmers disliked having seeds blow off their plants, so dispersal mechanisms were quickly eradicated during domestication. Dissemination of seeds is essential for spreading and establishing wild plant populations. Some Thai domesticated rice varieties retain their long awns used in plant reproduction. The same is true for domesticated plants: they must bloom at the same time to be harvested at the same time. Plants that have been domesticated are shorter, have more stems, and produce fewer seeds than their wild counterparts. Domestication, for example, eliminated the bitter and digestion-inhibiting properties of many SA-derived compounds. Humans preferred plants that were both safer and tastier.

Domestication's genetic underpinnings are now understood. The TGA1 gene causes teosinte's extremely tough kernel; however, a tga1 mutation was used during domestication to soften and reduce the size of the glumes in maize, making the kernels more manageable during harvest and consumption. Teosinte Branched1 (TB1), a gene involved in maize domestication, regulates apical meristem dominance. In contrast to teosinte, which produces numerous branches and inflorescences, maize plants with a TB1 mutation only produce a single stalk (Wardle, D. A. 2013). As a result, maize devotes all of its energy to developing a single ear of corn, unlike teosinte, which goes into a massive vegetative state. Jaenicke-Després and colleagues discovered three genes involved in domestication after analyzing modern and ancient DNA from this lineage and traditional maize varieties from Mexico in 2003. Teosinte's genes were much more diverse than those of modern maize. Scientists could piece together how the crop was domesticated by comparing modern maize varieties to their ancient DNA. The original wild ancestor had multiple alleles at a single locus, allowing humans to select the most advantageous traits.

2.4 PLANT MATERIALITY: THE RELATIONSHIP BETWEEN PLANTS AND HUMAN

Archaeologists are increasingly using the term "materiality" to investigate the relationships between people and artifacts rather than focusing on what artifacts "mean." From an early emphasis on object agency or the recognition of "material and nonhuman agency," the precise meaning and application of the term "materiality" have shifted over time to emphasize relationships, networks, and meshworks rather than the individual.

The agent must be someone or something. As people try to put more distance between themselves and animal life, animals are becoming more of a focal point in conversations. Plants have not played a significant role in the anthropocentric methods of zooarchaeology thus far. This paper will discuss how plants can be considered material objects and demonstrate their importance to society. We use a deliberately broad definition of "materiality," taking into account the material qualities of plants and the material and social ramifications of plant-human relationships. Following a brief introduction to "what plants want," the various facets of plant-human connections are discussed under an embodiment, investment, praxis, and entanglement headings. (Van der Veen, M. 2014).

What do plants require?

Plants are critical to life on Earth's survival because they provide food and oxygen. Plants provide us with more than just oxygen to breathe; they also provide us with heat, shelter, food, fiber for clothing, and beauty in the form of fruits, vegetables, and flowers. It is instructive to consider the following in light of Mitchell's research.

In a nutshell, what do plants genuinely require? The answer is astonishingly simple. Plants' primary motivation stems from their desire to reproduce. For this purpose, numerous strategies have been developed. Some (many) trees, for example, produce a lot of pollen and rely on the wind to spread it to increase their pollination chances. Some (many) flowers attract insects that aid in pollen dispersal, reducing the amount of pollen required.

On the other hand, some cereals do not require external pollination, and many other plants, including weeds, can reproduce from root fragments. Others have created spines that cling to the fur of passing animals or wind-borne aerodynamic seeds. Small animals' diets contribute to the spread of certain plant species because they consume the seeds or nuts of these plants. Plants produce phytochemicals, also known as "secondary compounds," in order to repel herbivores and attract beneficial insects such as pollinators (e.g., pigments). Several plant compounds and materials have recently gained popularity (poison, medicine, mind-altering substances, food, fibers, sweet smell, and beauty).

Some traits that evolved to discourage herbivores have been appropriated as pleasurable or desirable by some humans, demonstrating that traits like these can only exist in the context of specific relationships and networks (for example, herbs and spices). Because plants cannot flee predators, they must rely on these substances and strategies to survive. They cannot move to a new location to escape a predator or fertilize a nearby plant. As a result, they have devised strategies to use third-party resources such as humans, the wind, insects, birds, and other animals.

Many scientists have emphasized that humans and plants have historically co-evolved. Pollan (illustrates how domestication and selective breeding helped plants meet their primary need to reproduce by using the apple, tulip, marijuana, and potato. This allowed them to multiply and spread faster. Plants have discovered through trial and error that attracting animals, such as bees or humans, to spread their genes is the most efficient way for their species to reproduce (Van der Veen, M. 2014). Plants have spread worldwide because they satisfy our subconscious or conscious cravings for sugar, beauty, intoxication, and power. So, while we usually think of domestication as something humans did to other species, plants and animals may one day see it as "a clever evolutionary strategy to advance their interests." Wheat, potatoes, apples, tulips, sheep, and dogs are a few examples of "domesticated" species that have spread and become far more diverse than their wild ancestors.

Weeds, like crops, have formed mutually beneficial alliances with humans. Lycanthropes, or wild species that live alongside humans and benefit from our presence, include grain beetles, head lice, and rats. Despite the extensive use of herbicides, weeds still account for 10%-20% of arable productivity loss in today's farms. Most definitions of "weeds" center on plants that grow in unfavorable conditions, as observed by farmers and gardeners. On the other hand, a more holistic ecological perspective sees weeds as opportunistic species that thrive after human activity has disrupted their habitat and plants that compete with humans for soil possession. Anderson asserted that "the history of weeds is the history of man" and that the decision to cultivate a particular area also included the decision to weed. Weeds are highly valued in archaeological contexts because they are frequently found near crops. Annual/perennial life forms and ecological preferences for nutrient-rich/poor, acidic/neutral, and wet/dry soils can be used to infer cultivation methods, arable strategies (manuring, weeding, etc.), and production scales. Furthermore, these synanthropes demonstrate how problematic the nature/culture dichotomy is in Western thought. Weeds are not indigenous to the area but thrive in cultural or artificial settings. Their similar interactions with humans and human customs shape their shared experiences as "domesticated" species. To consider humans, crops, and weeds as all belonging to the same mutually constructed ontological realm and as unwanted intruders from "nature" is to miss the point entirely.

Weeds, too, employ a wide range of reproductive and survival mechanisms. Because annuals with short life cycles produce many seeds, they can quickly repopulate an area if they are not eradicated during the cereal crop's growing season. Some perennial weeds, such as bindweed, have become a nuisance in fields and gardens due to their ability to spread via underground rhizomes. By pursuing it where it begins, we contribute to its clever plot for global hegemony, as Pollan stated: "the evolution of bindweed accounted for the home." Wild oat, brome grass, darnel, and false flax are examples of "professional" weeds because their growth cycle, seed size, and shape are all similar to the crop, allowing their seeds to be harvested alongside the crop and resown alongside the crop at the start of the following season. Plants can meet an ever-increasing range of human needs, including food, flavor, beauty, intoxication, fibers, resins, dyes, timber, and shade. Plants use the needs of animals and humans to further their own goal of self-reproduction. The most attractive people have had tremendous success reproducing and spreading their genes. While there were times when both parties benefited, the connection had consequences for both people and plants. The term "coevolution" is commonly used in biology to describe this phenomenon, whereas "entanglement" and "relationality" are more commonly used in archaeology. (Van der Veen, M. 2014).

We are our diet

Plants are necessary for our diets because they provide the raw materials for all foods we consume, whether directly or indirectly. Plants, unlike other objects, are physically ingested into the body when consumed and thus incorporated and embedded in the body.

The study of this bodily substance is gaining popularity in archaeology. Food and its nutritional value may be plants' first and most direct influence on humans. When we do not eat enough of them or get enough of the proper nutrients, plant-based foods affect us physically and keep us going the same way animal-based foods do.

Plants' effects on humans vary greatly, with some having mild sedative properties that help us relax or sleep, while others keep us awake and alert, or even healing wounds, fusing bones, open airways, and fighting infections. Having enough high-quality food and being willing to share it is critical to our physical, mental, and social health.

The chemical footprint left by various food groups can be used to study the nutritional effects on the body. The bulk of "you are what you eat" archaeology research focuses on what was consumed (along with subsistence, agriculture, and trade-related aspects) and food's social and cultural significance. Stable isotope analysis examines chemical signatures left in bones by specific foods, whereas archaeobotany and zooarchaeology examine actual food preparation and consumption remnants (based on both plant and animal sources). Both investigate the cultural and social implications of eating, such as the unequal distribution of certain foods (such as meat and exotics) and how some people use food to express or cultivate a particular identity. Both are extremely useful due to the insights they provide into shifting subsistence patterns, agricultural practices, international trade, consumer preferences, and demographic variation. The emphasis, however, is unmistakably on human use of plants (and animals) or plant objectification.

The second impact category includes plants and plant-derived substances that are addictive or alter one's state of consciousness. Addictions can result from repeated use of stimulants and intoxicants such as alcohol, tobacco, cannabis, and opium, as well as drugs such as sugar and, to a lesser extent, tea and coffee. Each of these plant chemicals has the potential to alter a different aspect of human biology. This can result in a dependent relationship, which harms people's emotional, physical, and mental health. However, how can we tell the different plant substances apart? Westerners commonly classify plants as edible, restorative, psychoactive, or toxic, but this classification is not always accurate. While some substances can be therapeutic or pleasurable in small doses or concentrations, others can be toxic or addictive when consumed frequently or in large quantities. Treating the mind and body as separate entities is problematic, and even "nutritious" foods can affect a person's mental state. Although psychoactive substances are deeply embedded in human culture and social life, archaeological studies of their consumption are uncommon. In-depth studies of people's use of mind-altering substances and the contexts in which they were used are uncommon, but Lewis-Williams is an exception. For example, most studies of alcohol have focused on its role as a social lubricant and ritual participant. These concerns appear to have nothing to do with the "materiality" of relationships at first glance, but as we'll see, human cravings for specific foods and plant substances, as well as plant cravings for procreation, have resulted in intricate interactions between the two.

2.5 INVESTMENT – SEEDS AS ARCHIVES

Human interaction with plants has left a physical imprint on the plant's composition, resulting in a reverse embodiment. Two examples are agricultural practices and the effects of plant selection and mutation during crop dispersal. According to Fraser et al (2013)., evidence suggests that arable field manuring and irrigation, two standard soil enhancements, leave a chemical imprint on the seeds of crops grown in those fields. Manuring causes high nitrogen (15N) levels, whereas consistent carbon levels reflect the presence of water and irrigation. This situation benefits both people and plants. Crops can thrive and multiply in unfavorable or hostile environments if the growing conditions in the fields are optimized for their success. Expanding suitable habitats will also benefit invasive plant species and other plants. Increased or more consistent yields benefit people, and archaeologists gain valuable insight into past people's interactions with plants through the chemical signatures of crop seeds and the types of weeds that thrive in those fields (and are harvested alongside the crops). (Van der Veen, M. 2014).

Historical landrace crop DNA can tell us how and where certain crops spread beyond their native range, similar to how ancient seed DNA (if successfully extracted) can shed light on the interconnected roles of humans, plants, and the climate. Consider the following example:

Recent DNA analysis findings indicate that multiple domestication events occurred in the Near East for several crops, including emmer, einkorn, and barley, affecting our current understanding of intentionality. If this is the case, it will imply that barley was brought to Europe on multiple occasions. Today, Europe is home to three distinct but related types of barley. They are distinguished by their latitudinal distribution pattern and their origin in South-West Asia.

According to genetic and archaeological evidence, this particular barley variety was introduced to Europe much later, possibly in the fourth millennium BCE, along with other agricultural innovations. Previously, explanations for this phenomenon included the relative success of hunter-gatherer societies, cultural preferences for specific foods, and the need for crops to adapt to new climatic conditions. Although other possibilities cannot be ruled out, the new DNA evidence suggests that the latter explanation, namely the introduction of day-length non-responsive barley, was most likely a significant factor. Plant material properties help us understand the intricate connections between vegetation, weather, and human societies. Plant characteristics can change over time. Plants can be used as archives of past human and plant behavior because they adapt to human interactions.

2.6 PRAXIS – A WAY OF LIFE

Everyday plant-related actions, such as harvesting, caring for, cultivating, or growing plants, also form embodiment (Ingold, 1993). Cereal farmers and their daily, monthly, and yearly rhythms are mirrored by the cultivated crops.

Such farmers grow cereal crops. Soil preparation, for example, planting, weeding, harvesting, threshing, and storing, is a seasonal activity. The motions required to operate the tools of the trade (spade, plow, traction animals, threshing stick or sled, sieves) are also fixed. Their senses of self are shaped by farming and the yearly repetition of the same set of behaviors. They become farmers, a distinct way of being whose ontology is based on specific, historically derived relationships in flux rather than an eternal essence. The farmer's embodied routines shape how he or she experiences and interacts with his or her surroundings, which include the landscape, plants and animals, and other people. In eastern Turkey, a farmer and his family were harvesting their crop. When asked why he did not use a scythe and cradle to cut harvesting time in half, the farmer responded, "Then what would I do?" A farmer's relationship with his or her plants is fundamentally different from that of, say, a hunter-gatherer, as are the farmer's animals, tools, buildings, and communications. Though they share the responsibility of providing for their families and other necessities, men and women in material-culture societies organize their days and allocate their time in very different ways.

Instead, gardening and horticulture move at their own pace. Small plots of root and tuber crops, vegetables, herbs, fruits, and cereals are more common than large fields of one or two annual crops. Garden plots are typically located near residences because many of these plants thrive in intensive cultivation with frequent monitoring and care, utilizing the "little and often" strategy. As the plants become closer and the relationships between plants and people become more intimate, more personal care is required for each plant, such as planting, dividing, taking cuttings, weeding, watering, and feeding. This closer relationship between cultivator and plant may have resulted in increased experimentation, the introduction of new plant varieties, and the importance of plants in human ritual, medicine, and other pursuits. The proximity of the garden plot to the home may reinforce the notion that gardening is primarily a "domestic" activity performed by women. Caring for plants daily is analogous to caring for a baby or young child. According to ethnographic research, women are more likely to be involved in hoe-based cultivation (gardening, horticulture) than men are in plow-based agriculture (field crops). This emphasizes the differences in involvement and timing between mothering/nurturing and husbandry.

Each harvest will have its cadence. During the Islamic era, many so-called "summer crops," such as citrus fruits, sugar cane, cotton, rice, and bananas, were introduced to the Middle East, North Africa, and Spain (AD 700–1200; Van der Veen, 2011). Plants from South and Southeast Asia and the Pacific require warm temperatures, plenty of sunlight, and consistent watering to thrive in cultivation. They needed to be irrigated in North Africa and the Middle East because summer rainfall was insufficient. This not only necessitated significant infrastructure investment (digging and maintaining irrigation canals, building and maintaining water wheels, and arranging access to available water; fertilizer) but also significantly altered the agricultural year and, by extension, the farmers' lives. Cereals, pulses, and oil/fiber crops were traditionally planted in the fall and harvested in the spring in regions with mild winters and warm summers. Summer months have historically been slow in the agricultural sector, except for some irrigated fruit and vegetable farming. By the tenth or eleventh century, extensive irrigation networks had been established. In the tenth century, al-Masudi described Egyptians as "the richest men in sugar," while al-Idrisi described the area around Cairo as "a fertile field of sugar cane" in the twelfth. As a result of year-round agricultural activity, the workload and farming methods have increased dramatically. Forget about relaxing during the summer. Large-scale irrigation is often associated with large-scale land ownership, and several new crops are well-suited to plantation or "industrial" farming techniques. Because of the new yearly cycles, embodied technologies, and labor relations, life will be different (Van der Veen, 2011). Fruit tree cultivation is difficult for a variety of reasons. Trees can take up to a human generation to bear fruit and live for decades. Fresh fruit and vegetables arrive every year, but unlike cereals, they are only sometimes suitable for long-term storage. Gardeners who plant apple or plum trees are probably accustomed to an annual fruit glut. Farmers in this situation face long-term planning, financial investment, and a wide range of tree-growing practices. This is why orchards and market gardens are frequently located near population centers such as towns. They must also consider where they will sell their wares. Farmers growing cereals and vegetables, gardeners, sugar cane landowners, and horticulturists growing fruit for markets all have a direct relationship between the materiality of their crops and their sense of reality. Their very survival depends on their plant life's quirks, cycles, and requirements. As a result, distinguishing between agricultural and hunter-gatherer societies is obsolete.

2.7 ENTANGLEMENT – WEBS OF RELATIONALITY

So far, interactions between plants and humans have been described as mutually beneficial, co-evolved, co-dependent (biologists' terms), or as a source of cultural innovation. Unlike "network," which implies connections between discrete nodes, Ingold's (2007) term "meshwork" describes the intertwined lines along which materials flow, mix, and mutate. "Humans get caught in a double bind, relying on things [plants] that rely on humans," writes Hodder (2011, 164) to describe the tangled or trapped nature of plant-human relationships. The transition to agricultural life is an example of this complication. For example, Harlan (1992) recognizes that cereal crops have "domesticated" humans as a result of our reliance on them and thus classifies them as "artifacts." A small percentage of the human population can survive without cultivated plants, but a fully domesticated plant cannot.

Humans and plants are mutually dependent, so one cannot exist without the other (Harlan, 1992). Harlan discusses how domestication has made cereals and pulses reliant on human dispersal for reproduction due to the loss of natural dispersal mechanisms (loss of the brittle rachis or pod dehiscence at maturity). A single mutation can cause natural seed dispersal loss in cereals, and selection for those mutants may have occurred unintentionally, for example, due to a change in how wild cereals were harvested. As a result, people may have unintentionally become more attached to plants and become entangled in a web of labor-intensive activities (Fuller, 2010).

Farmers became dependent on crops because they have material properties—they must be planted in one location until they mature, produce fruit at a specific time of year, and require care and attention before, during, and after the growing season (though Harlan, of course, never used that term). Agriculture does not necessitate sedentism; groups can cultivate small plots of land while remaining primarily nomadic.

Long-term storage of cereal grain (and other crops such as pulses) paves the way for permanent settlement and the production of a food surplus that can be used to alleviate scarcity and allow for periodic feasting and gift-giving. This is critical in the new agricultural era. Land ownership plays a vital role in social relations within industrial farming systems, hastening the emergence of new asymmetrical modes of existence, such as landlord and tenant farmers, as well as the accumulation of wealth and social inequality. Benefits would have included a more consistent harvest, better control over available resources, and the ability to propagate the plant beyond its native range. Because of changes in the flora, fauna, landscape, material culture, and social structures, it was often impossible to return to pre-change ways of life. (Robb 2013). Agriculture's development is a prime example of human interaction with plants, as it paved the way for sedentism, property rights, and wealth accumulation. These changes had an impact on many different aspects of human life, the natural world, and the physical universe.

This type of tangle is also visible when people introduce new crops into different areas. Summer crop adoption in the Middle East is a classic example. Political stability, increased mobility, changes in land tenure, economic growth, rapid urbanization, and land use were all intertwined with the development of Islamic culture and the exceptional receptivity of early Muslim civilization (entanglements or meshworks, as we would now say). Many of the new crops had the potential to become cash crops, and farmers in the areas surrounding emerging cities could benefit from this. While proximity to new markets or irrigation water enabled one to benefit from the new prosperity, it also increased one's tax exposure and reliance on those markets. The local environment, socioeconomic changes, opportunities, dependencies, inequalities, and increased regional diversity all influenced the plants' unique characteristics.

The history of sugar shows that its influence and connections did not end with its introduction to the Middle East. This paved the way for sugar consumption to spread rapidly throughout Europe. Sugar was introduced to the Caribbean as an exotic and costly luxury item, prompting the establishment of extensive plantations to meet local and regional demand. Because of the labor required to harvest the crop, slavery was necessary to reduce its cost. European-introduced diseases wiped out the indigenous population, and enslaved Africans from the transatlantic slave trade replaced them. While destroying the region's rainforests, enslaved Africans drove down the price of sugar, making it profitable to grow on new plantations. They rose to prominence in the Caribbean and eventually ruled the region. This production method reduced the sugar cost to the point where it could be mixed with tea, providing British factory owners with the ideal break-time "food" (energy plus stimulant) to keep their workers going strong during long shifts. Sugar impacts modern society despite its widespread availability and addictive properties. Aside from providing a reliable and nutritious staple and promoting population growth, the introduction of American maize into West Africa may have provided additional African labor for American sugar, cotton, and tobacco plantations. Other plants that have had a long-term impact on human society include tea, cotton, coffee, cocoa, and potatoes, as well as quinine, which allowed Europeans to colonize parts of Africa and Asia by reducing the effects of malaria.

Emotional responses were a notable side effect of newly introduced food plants, and emotional archaeology is a relatively new field of study (e.g., Gosden 2014). Because of the links between food, emotions, and interpersonal relationships, plant-based foods significantly impact our social environments. Novel foods, like collections of similar items, can spark the emergence of new social entities and connections. For example, the Roman conquest of northwest Europe introduced Mediterranean foods to the region, which may have been associated with a different culture. Fifty new foods were introduced to Roman Britain, including spices such as black pepper, Mediterranean fruits and vegetables such as grapes and olives, and "English" staples such as apples, plums, and cabbage. Their physicality and otherness would have elicited emotions ranging from curiosity and pleasure at the novel flavors to apathy, disgust, aversion, or economic and social opportunity. People will initially associate new foods with the new government, which will have social consequences. People who cannot afford the new, expensive foods or are unsure how to prepare them may feel excluded, inferior, or inadequate. Some may have encountered hostility, with people opposing anything related to the alien culture. Although these novel foods were initially associated with other cultures, this was not always the case. Some quickly settled in and assimilated into the local food and farming systems. The introduction and physical presence of these new foods prompted people to make decisions and form new alliances, which influenced the changing social and material conditions in Roman Britain.

This manifests plants' "materiality," which has far-reaching implications for human relationships, economic activity, and food consumption. Human decisions and treatments impact the plants' physical and mental well-being. The ability of people to mutate and thus thrive in new environments, combined with their desire for food, beauty, and other qualities, has resulted in complex and varied webs of relationships. While the plants have lost some of their independence and ability to reproduce (at least for some crops), they have gained a higher reproduction rate and a much wider geographic distribution than their wild ancestors. Humans have also benefited from the cultivation of fodder crops, as it has provided them with a wider variety of foods, flavors, plant fibers, nutrients, and animal products. They had less freedom, were less active, owned, socially stratified, were forced to eat certain foods, or were dependent on exploitative production methods; they also lived in larger communities. Although humans may have been willing slaves in the plant's reproductive processes at one time, some crops enslaved humans.

2.8 CURRENT USES OF PLANTS

Plants are necessary for a healthy ecosystem to function. Several species can coexist peacefully on the bark of a single tree. On a microscopic level, each plant or tree is a self-contained ecosystem home to a diverse range of organisms. Plants are essential to everything from ants to elephants to humans. (Van der Veen, M. 2014).

Food sources

Plants are the primary source of food in nature. Everything we eat or drink is related to the plant kingdom in some way. Due to their diversity, we cannot categorize plant-based foods into a single category. Plant-based foods are divided into four major categories to make them easier to understand.

- Direct Foods: This category includes all fruits and vegetables, whether raw or cooked. Mango, guava, pumpkin, gourds, melons, okra, eggplant, and many other fruits and vegetables fall under here.
- Edible Oils: Every oil type we use for cooking comes from plant sources. Plants provide rice bran oil, canola oil, castor oil, olive oil, mustard oil, and palm oil.

- Indirect Foods: This category includes plant-based processed foods and plant-derived products. These foods are not commonly found in nature. Plants are the source of many food items such as sugar, chocolate, jams, and juices.
- Food flavorings: This category includes foods we do not normally consume directly. Instead, these items are used to either enhance flavor or create new tastes. These plant-based ingredients can also be used to season food. Spices in this category include ginger, cardamom, cinnamon, and even vanilla. Herbs like mint, fennel, and basil can also be included in this category. These are all derived directly from plants.

Shelter

Our improvised shelters are made of trees and plants. Since ancient times, our forefathers have lived in caves and jungles. They lived on grass beds beneath the canopy of trees. They eat meat and fruit for food. The origins of humans can be traced back to forests and water sources. This support system was ideal for humans at the time. However, do we still need plants for shelter? Consider that every door, window, and piece of furniture we own is made of plants. This is the fundamental truth of human society.

Today, we use more wood in our homes and shelters than our forefathers did 100,000 years ago. The only thing that has changed is how well-organized we are now when using wood, particularly in our homes. (Van der Veen, M. 2014).

Source of medicine

Ayurveda, the traditional medical system, is entirely based on plant sources. According to ancient Indian researchers, various plants were used as medicines as far back as 5000-10,000 years ago. Many of these records have been confirmed to some extent by modern science to be accurate and valid today. In modern medicine, many medicines are made from plants or products derived from plants. Other medicinal plants, such as tulsi, neem, giloy, and ashwagandha, are now widely used.

Source of cloths

Some may argue that our fabrics contain the majority of modern synthetic fibers. So, how can we rely on plant-based cloth? Our question has a simple, straightforward answer. Plants are the raw material for all fibers, natural and synthetic. Cotton is still the most popular natural fiber in the textile industry.

Source of energy

The use of firewood is steadily declining. Wood, on the other hand, remains a significant source of cooking fuel in developing countries. We may have used some as fuel if we enjoy grilling and cooking. Firewood is a primary source of fuel in rural areas. It was the only source of fuel until the early nineteenth century. Wood is also used as fuel in some commercial and industrial projects.

Paper and Rubber

Paper and rubber are two of the most important inventions in human history. Our understanding of history depends on the invention and development of paper made from bamboo leaves in Ancient China and India.

On the other hand, rubber has changed how the transportation system works, from soft and adaptable modern tires to bouncy hardwood cartwheels. All of this was only made possible by the discovery of natural rubber. Natural rubber is still widely used in the latex industry.

Source of perfumes

Herbs and spices were the primary sources of fragrances and aromas in antiquity. However, there was a disadvantage: spices have a strong odor. In order to achieve a more pleasant, mild, and sweet scent, People have begun extracting concentrated oil to create desired fragrance notes. Plants are most beneficial when used in perfume. Rose, sandal, jasmine, and cardamom oils are still very popular in the perfume industry.

Pollution control and air purification

Plants are the most effective way for nature to reduce air pollution. It is the primary function of all plants. Remember, A larger-sized plant can filter more air per day. In nature, plants maintain the balance of oxygen and carbon. Although carbon dioxide is a more critical gas than oxygen, we usually consider oxygen to be the most important. This is the essence of existence. It is necessary for plant survival. Finally, we rely on the carbon cycle. Humans, unlike plants, inhale and exhale carbon dioxide. Plants absorb carbon dioxide and expel oxygen. Humans and plants have a close symbiotic relationship.

Temperature control

Forests serve as the planet's air conditioning system. Plant moisture helps to moderate air temperature, so this is a reasonable assumption. On a global scale, thermal exchange regulates climate and temperature.

This is especially noticeable on hot summer days. Walk around the lawn for a few minutes. Then, find some shade under a tree. The temperature will rise dramatically. Plants regulate their internal temperature in this manner. This could be one of the reasons why plants are so soothing to the eyes and mind.

Soil protection

You may have noticed soil being washed away by flowing water on a rainy day. Soil erosion is the name given to this phenomenon. It does not appear to be dangerous or concerning on a small scale. However, when observed on a global scale, the threat to human life becomes clear. All crops and food plants are nourished and sustained by this topsoil. No plant could survive if the soil were eroded entirely. To prevent soil erosion, more trees must be planted. The thick and fibrous roots of some plants hold the soil firmly in place and prevent soil erosion caused by moving water.

Furthermore, the plant can help to maintain soil fertility. The decaying and decomposing leaves and branches enrich the topsoil with manure and organic fertilizers. It also benefits plants and other animals by storing water. The turgor pressure of the plants helps to raise the groundwater level. One of the most severe issues in many developing and developed countries is a lack of access to underground water.

Chapter

3

PURE AND APPLIED BOTANY

Contents

3.1 INTRODUCTION

Botany is one of the several fields in biology. It involves the study of plants while looking into their biochemical processes, their structure and properties. On its own, botany is very wide such that it is further subdivided into two groups namely pure botany and applied botany. Botany covers the classification of plants. It also seems to bring an understanding of the various plant diseases and how plants interact with their surroundings. Botany has brought to light various information and principles which have formed the basis of other fields such as applied sciences such as forestry, horticulture and agriculture. Botany is very broad and this is attributed to the fact that plants are very important to humans more so the early humans as they used plants as medicine, food, clothing, shelter and tools among others. Plants have both economical and practical values and are therefore indispensable to all life on earth. Various processes occur in plants that facilitate the sustenance of the environment. They include processes such as photosynthesis which allows the plant to generate its own energy. The process is said to make life possible. As mentioned, there are two main classifications of botany including applied botany and pure botany. They cover different parts of botany. (Charaya, M. U. 2020).

3.2: PURE BOTANY

Pure botany is under botany in general and pertains to the study of plants all of which form part of nature. There are 10 different classifications of pure botany. They are discussed as follows:

3.2.1: Morphology

Morphology as a part of pure botany entails the study of the structure and form of plants. Plants are made up of various plants including roots, fruits, leaves, stems and flowers. They are all studied under plant morphology. It is in plant morphology that we discover that there are plants that will have flowers while others will not. Under plant morphology, there are various subdivisions including histology, concerned with the study of tissues, cytology, concerned with the study of cells, reproductive morphology which involves the study of life cycles, anatomy covers the internal elements of the plants including the internal organs of the plant and morphogenesis that looks into the development of the plant. Plant morphology can generally be defined as the study of the structure, form and development of plants. It seems to give more information on the structure, form and development of a plant on the basis of the similarity of origin and plan. (Vyshivkin, D. D. 2017).

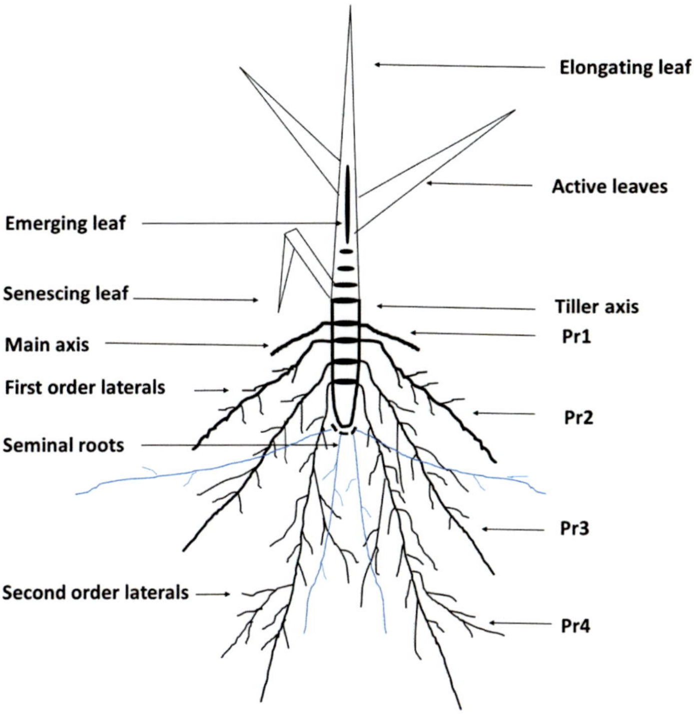

Figure 3.1: Developmental stages of leaf and root-bearing phytomers (Pr) at the tiller axis of wheat plant in a stylized diagram. Pr, root-bearing phytomer. Blue lines represent seminal roots.

Source: Robin AHK, Ghosh S, Shahed MA. PEG-Induced Osmotic Stress Alters Root Morphology and Root Hair Traits in Wheat Genotypes. *Plants*. 2021; 10(6):1042. https://doi.org/10.3390/plants10061042

Under the scope of plant morphology, there are four main areas where the studies are focused on. Each of the four areas tends to overlap with each other in biological sciences as a field. Plant morphology is very unique. There are certain elements that can be used to define plant morphology. They include the fact that on its own, it tends to be very comparative. This is attributed to the fact that plant morphology gives the structure of plants and makes comparisons on the different kinds of plants. For this reason, morphologists may tend to use various plants in their studies and make comparisons between them. From the generated data, morphologists are able to formulate ideas on the various plants. The studies have brought about terms such as homologous. The term is used to refer to structures in species of plants with differences brought about by their inherited genetic pathways. We can take the case of cabbages, oaks, and pines known to have similar leaves though they look very different. All these plants are said to have similar basic structures and similarities in the arrangement of their parts. Morphologists may agree that the discovery of homologous leaves was quite easy. The spines in cactus plants have also been found to be homologous as they have the same development patterns and structures. Such kinds of studies in plant morphology therefore may overlap the study of paleobotany and plant evolution.

Plant morphology can also be identified with the fact that it entails the observation of the reproductive and vegetative structures of plants. Studies show that plants have different reproductive structures. The kind of reproductive organs found in plants are said to be specific to certain groups of plants such as the fern sori, flowers, moss capsules and seeds. In the case of vegetative structures, studies are mostly done on the roots of the plant, the leaves, stems and the shoot system. The various studies on the vegetative and reproductive structures have led to various discoveries. For instance, the study on reproductive structures led to the discovery of existing alternations in the generations of various plants including algae. In this sense, studies on vegetative and reproductive structures may tend to overlap with the study of plant systematics and biodiversity. (Charaya, M. U. 2020).

Plant morphology also looks into the study of plant structures within certain scales. There are various ranges to the scales. For instance, the smallest scales are in the ultrastructure. This scale is employed when conducting studies on the basic structural features of cells. An electron microscope is among the very few devices that can be used in the study of such structures. The other scale is involved in cytology often utilized when optical microscopy is used in studying cells. When using such kinds of scales there is a sense of overlapping with plant anatomy. In the field of botany, the largest scale is used when studying the growth habits of plants while taking into consideration the general architecture of the plant. It may be noted that among the various species of plants there is a variation in the pattern branching in trees. The same could be noted with regard to the general appearance of grass, herbs and trees.

The other defining factor of plant morphology is that it is used in examining development patterns in plants and studies the origin of plants.

Under this category, morphology looks into the various processes through which plant structures originate and mature throughout the growth process. Plants are very different from animals in that animals as they are born, have various parts during their growth and continue with those parts till they die. This is not the case in plants as they are known to produce new structures and tissues throughout their life. For instance, embryonic tissues have been found in living plants. Studies show that the generation and growth of new plant structures are dependent on the plant's development stage. Another factor affecting such an environment is the kind of environment the plant lives in. For instance, if there is a deficiency in water supply then the plant may have to develop certain features to enable them to grow or when the plant has attained the stage of reproduction, it will develop the reproductive organs needed for the new stage. This part of plant morphology tends to overlap with the ecology and physiology of plants. (Shipunov, A. 2018).

As mentioned earlier, the morphology of the plant looks into both the vegetative and reproductive structures of the plant. While looking into the reproductive structures of vascular plants, there are two main classifications characterized by two main systems. They include the root systems and shoot systems. The shoot system is made of leaves and stems. Both systems have been found in almost all vascular plants hence creating a theme of uniformity in plant morphology. This is not the case in plant reproduction as different plants have different reproductive structures. Some reproductive organs are common to certain plants. For instance in flowering plants, it is common to find both flowers and fruits in angiosperms, seed cones are found in gymnosperms and conifers while sori are found in sperms. For this reason, the vegetative structures of plants can be used in distinguishing plants from one another.

During studies and analysis, it was found that the various plant morphological characteristics are comparable, measurable, countable and describable and can be used to show the similarities and differences in plant taxa. The unique characteristics of plants are used in identifying, classifying and describing the plants. The characteristics are termed as key or diagnostic when they are used in the process of describing and identifying the plant. Plant characteristics can be either quantitative or qualitative in nature. The qualitative characteristics are morphologic in nature and give information on the pubescence, flower color and leaf shape while the quantitative characters are plant morphological features that can be measured or counted. For instance, certain plants are known for the number of petals they produce. Both the quantitative and qualitative features of plants are used in identifying plants. (Vyshivkin, D. D. 2017). During the various studies on plant vegetative structures, the generated evidence led to the discovery of the occurrence of alternation of generations. The alternation of generations was found to occur in several plants and certain species of algae. This discovery is said to form the basis of understanding plant life cycles. Morphology also looks into the major function of pigments in plants that is photosynthesis. The process is known to make use of chlorophyll which is the green pigment. In some cases, the process also makes use of yellow and red pigments to allow optimal capturing of light. Researchers credit the pigments to be a very important factor in drawing insects to flowers to facilitate pollination.

The pigments found in plants are composed of various kinds of molecules including the betalains, anthocyanins, carotenoids and porphyrins. They are all well adapted such that they have the ability to selectively absorb certain lift wavelengths. This also allows the reflection of other wavelengths. Chemical reactions in plants make use of the absorbed light. The color of the pigment can be determined using the reflected wavelengths of light.

3.2.2: Histology

In pure botany, histology entails the studying of plant tissues and structures. Studies of plant tissues show that the structures of plant tissues have direct relations to their functions. This means that there is an overlap between histology, physiology and anatomy. Under histology, there is the term histopathy commonly used to refer to the study of plant tissues that can be affected by diseases. Histopathy is very useful in making a diagnosis of a plant, the stage of plant diseases and the rate at which the disease is spreading and affecting the plant. As the disease affects the plant, the processes are known to affect plant tissues in various ways. Effects of various diseases on plants are affected by the type of disease, type of tissue and the level of disease contamination.

When measuring diseases in plants, researchers make use of units called the histopathology units. The units can be found in several hospitals and some independent private laboratories. These laboratories offer services that can be accessed by healthcare practitioners. Histopathy makes use of plenty of tests most of which are readily available. These tests are usually conducted by highly skilled practitioners. Their skill will enable them to understand and interpret the results of the study. In most cases, the various laboratories may specialize in the study of certain diseases. For instance, studies on the plant nervous system is conducted by the neuropsychology laboratory. (Charaya, M. U. 2020).

From the information so far, we get the sense that histology greatly involves extensive biomedical research that will provide the necessary information on the causes and possible medical treatments for the identified disease. Research studies done under histology may in most cases be done in a hospital laboratory. There are cases where the studies are done in pharmaceutical companies, research institutes and universities. Under histology, the term histopathologist is very common as it is used to refer to people with the ability to look down a microscope and come up with findings that they can interpret. They are very different from pathologists in the sense that the pathologist does not depend on the microscope to get information on the disease affecting the plant. Histology makes use of very tiny slices of the specimen that are presented before a microscope. Some information can be obtained by just mere views go the plant while others may require the specimen to be put under a microscope to generate the important details. For the latter, it is very necessary to make use of surgical removal procedures as the processes rewire very thin views of the specimen. There are cases where a biopsy is to be done. In such a case the required cells are extracted at a particular time and placed in a microscope. There are various microscopes used in histology. They include a light microscope and an electron beam microscope. The light microscope is used to obtain the basic features of the cells. There is also a virtual microscope. As the name says, it is virtual and works by mirroring the functions of a real microscope. (Shipunov, A. 2018).

Those using the microscope or an electron microscope have to be very familiar with its working. They need to be familiar with the parts of the plants such that when they study them, they are able to know what is off in the plant. Histology is also comparative as it involves making comparisons between a very healthy plant cell and the diseased plant cell. Virtual microscopes can be found on websites online. They can be found on desktop computers, tablets and laptops. (Vyshivkin, D. D. 2017).

3.2.3: Physiology

Physiology is a branch of pure botany that deals with the study of plant functions. Initially, it was classified under other branches of pure botany. It was greatly linked to the development of other components of pure botany including morphology. This is attributed to the fact that there is a great interrelationship between the structure of the plant and its functions. Studies show that they are very closely related such that it is hard to distinguish one from another. While studying the functions of plants, it may be necessary to get information on the different aspects of plant structures. This is very key to understanding the roles. It is also important to note that gradually over the years, there has been an evolvement in plant functions, structure and forms. There is a great interrelationship between physiology and other fields of biophysics and biochemistry. This is attributed to the growth and evolvement of methods used in studies. They have been very key in shedding light on various aspects of plant physiology.

3.2.4: Ecology

Ecology as a subdivision of pure botany entails the study of mutual relationships between existing plants and their physical environment. There are different kinds of relationships ranging from saprophytic, xerophytic and hydrophilic. The physical environment is composed of a number of factors including the climate of the area, soils used in plant growth and other conditions. In a basic sense, plant ecology is described as a phase of plant physiology present in either controlled or uncontrolled environments. Ecological studies of plants show that they are very sensitive to the forces of the environment. They are also affected by their geographic distributions and their association with communities most of which are greatly determined by soil and climate character.

3.2.5: Anatomy

Plant anatomy also known as phytotomy is a term used to refer to the study of plant's internal structure. In the early stages of pure botany, plant anatomy was studied under plant morphology thereby describing the internal and external structures of plants. Placing them both together makes the study very broad and so plant anatomy was later on classified as a field on its own. It places much emphasis on the study of the internal structures of plants. Plant anatomy has grown over the years such that it now allows the study of plants at the cellular level. Some of the major components of plant morphology include microscopy and sectioning of plant tissues. (Shipunov, A. 2018).

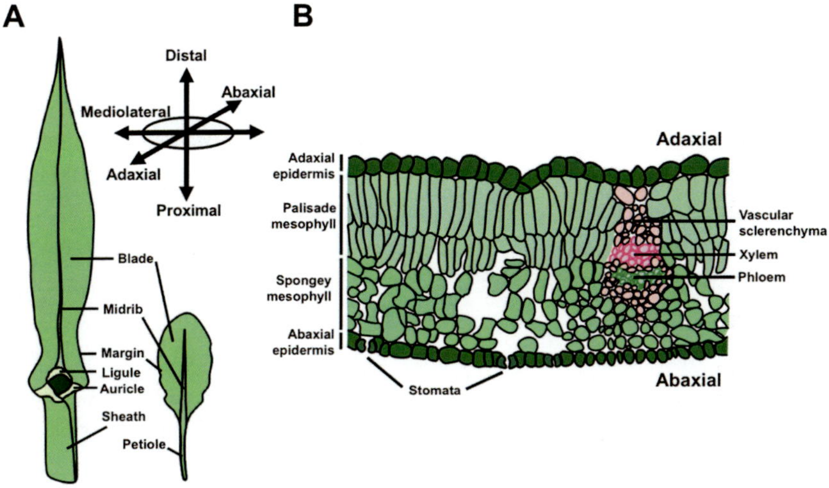

Figure 3.2: The structure of a leaf: (**A**) Leaves from maize, a monocot, and the two closely related eudicots, *Cardamine hirsuta* and *Arabidopsis thaliana*, highlighting major features of leaf anatomy. (**B**) Cross-section of a mature eudicot leaf illustrating the characteristic features of the adaxial and abaxial leaf domains.

Source: Satterlee JW, Scanlon MJ. Coordination of Leaf Development Across Developmental Axes. *Plants*. 2019; 8(10):433. https://doi.org/10.3390/plants8100433

Studies of plant morphology make use of various systems. The systems approach of plant studies is usually organized on the foundation of plant activities such as pollination, flowering, and transportation of nutrients, seed development and embryogenesis. The systems approach has been classified into various structural categories including fruit and seed anatomy which deals with the study of the structure of the accessory fruit, pericarp, seed and ovule. The stem anatomy entails the study of the shoot apex, buds, vascular tissues and stem structures. There is leaf anatomy which includes the study of plant palisade cells, stomata and epidermis. Root anatomy involves the study of the gynoecium, androecium, corolla and calyx. Wood anatomy deals with the study of the branches, collar, sapwood, heartwood, vascular cambium, phloem, xylem and bark. Finally, there is the root anatomy which deals with the study of the endodermis, roots and root tips (Charaya, M. U. 2020).

3.2.6: Geography

Plant geography deals with the study of the spatial distribution of vegetation and plants. Studies show that the various factors of the environment may affect the distribution of plants and vegetation. There are various words used to refer to plant geography that includes vegetation science, geographical botany, geobotany, phytochorology and phytogeography. Plant geography pertains to the aspects of plant distribution. It looks into the control of the distribution of different animal species both small and large species of plants and all the factors influencing the composition of floras and entire communities. In as much as geobotany is used to refer to plant geography, they are a bit different in the sense that geobotany places much emphasis on the effects of geographic space on plants. Generally, plant geography falls under biogeography. Those involved in the study of plant geography are known as phytogeographers. Their work entails the study of the processes and patterns in the distribution of plants.

There are four main fields of studies in plant geography. The fields include plant origin, plant community, plant taxa and plant environment. The four bring about the relationship between plant geography and other subdivisions in pure botany. For instance plant ecology and plant geography come together in the sense that there is an association between biome study and vegetation. Plant geography is further subdivided into two main categories namely historical phytogeography and ecological phytogeography. Plant distribution is affected by biotic and abiotic interactions. The abiotic factors are concerned with the reconstruction of the origin, distribution and extinction of taxa. Plant geography is heavily dependent on occurrence records that show and prove the presence or absence of a certain species of plant at a particular point in time. The occurrence records make use of operational geographic units such as the geographical coordinates or the political units. Data from these records are then used in the construction of phytogeographic provinces. (Vyshivkin, D. D. 2017).

Plant geography seeks to answer certain questions some of which are shared with other fields such as zoogeography. Zoogeography is a field concerned with the study of animal distribution. As the term phytogeography is very broad, its utilization is mostly done as per the manner in which periodicals use them. Some of the topics covered in plant geography include the distribution of plant species, historic biogeography, phytogeography and the distribution of genetic variation. In as much as plant geography seeks to create an understanding of plant distribution patterns, it does not place much emphasis on biodiversity patterns. When a group of plant species occurs in a given period of time they are collectively known as flora. Each of the plant species categorized under the flora has independent relationships and abundance with other species. The different flora can be formed on the basis of floral elements. Such elements can be as a result of the genetic traits of the plants. This means that the plant species have similar genetic information such as their evolutionary origin. The evolutionary element gives more information on the common plant route access to certain habitats. They may also have similar historical elements in that those species of plants experienced similar events. The similarities could also he as a result of the ecological element as they are found living in environments with the same climatic factors.

Such studies bring about terms such as population which is a term used to refer to a collection of all interacting individuals of certain plant species in a given area. In botany, an area is defined as the location in which plant species, plant elements or flora can exist. This brings about the term Aerography which refers to the study of the description of a given area and chorology defines the study of the development of plants. The topography of a region is given by the distribution of plants within that region. The study of the area is very important as it sheds light on how plant species interact. The area's margin, general shape, size and continuity relative to other areas have been useful in giving more information on plant geography. For instance, the term vicarious is used to refer to mutually exclusive plants. The surface of the earth can be divided into various groups including the floristic region. The floristic region is made up of various regions all of which have distinctive flora.

3.2.7: Genetics

Plant genetics similar to human genetics entails the study of plant genes, genetic variations and heredity. Plant genetics is covered in botany and biology. It is also interrelated with other life sciences more so the studies pertaining information systems. Comparisons made between plant genetics and animal genetics show that there are great similarities between the two while variations are noted in certain areas. Similar to animals, plants use their DNA to pass their traits to other generations of plants. One of the variations between plant and animal genetics is the fact that animal genetics places much emphasis on lineage and parentage while this may not be the case for plants as some plants are naturally self-sterile. Studies on plants show that speciation is very necessary as various groups of plants exhibit different genetic abilities including their ability to survive different environments. Plants are very different from other organisms as they possess the ability to produce energy-dense carbohydrates. This is achieved through a process known as photosynthesis. Plant chloroplasts are involved in this process. They also have their own DNA. For this reason, chloroplast can be considered to be an additional reservoir for genetic diversity and genes.

Information on plant genetics has some economic effects. Studies on plant genetics have been used in facilitating genetic modification of plants to lead to an increase in yields, reduce pest and disease incidence and increase the amount of nutrients present in plants. Some studies on plant genetics are specific to certain plants. It looks into the transmission of traits from one plant generation to another through the use of DNA. As mentioned earlier, chloroplasts carry some DNA that contributes to plant characteristics. Naturally, plant experience some somatic mutations on a regular basis. The occurrence of such mutations has been said to affect plant germ lines. This allows the development of certain components including flowers most of which develop at the ends of branches. For these reasons, most plant ends are composed of somatic cells. Such branches are called sports. New cultivars are obtained from fruits in sports having economically desired features. (Vyshivkin, D. D. 2017).

Studies on plant genetics show that certain plant species can survive on their own as they reproduce through self-fertilization. This rarely occurs in animals. This feature is very useful as it ensures the continued existence of the plant species. However, some scientists are looking into the possibility of crossing certain plant species to fully eliminate self-fertilization. This brings about the term plant breeding which allows the creation of new plant hybrids bearing the required characteristics. For instance, plant hybrids tend to have higher yields compared to the original ones. Other uses of plant genetics include the prediction of plant combinations that can be used to produce plants with high hybrid vigor. The studies have also been useful in providing information on the effects of hybridization. Plant genetics interrelates with other fields such as plant pathology. Plant pathology is defined as the study of plant diseases. Some plant diseases are caused by environmental conditions or pathogens. Plant genetics is useful in giving more information on the likelihood of certain plants being susceptible to some of these diseases.

This means that plant genetics could be used to provide more information on how such diseases can be prevented or treated. This goes a long way in ensuring that there is reliable food production. It also helps farmers tackle some agricultural problems. Plant genetics proves that cultivated and naturally occurring plants tend to carry some resistance to some diseases. Information on plant genetics also enables farmers to easily identify the most suitable areas or soils to be used in planting. Based on the genetic composition of the plant, it is easy to identify certain drugs that can be used to treat various diseases. It also sheds light on the most successful chemicals that could facilitate fast growth in plants and increased production. (Charaya, M. U. 2020).

3.2.8: Taxonomy

The term taxonomy is used to refer to the scientific classification of organisms. This means that plant taxonomy refers to the scientific classification of plants. The term classification is defined as the grouping of plants on the basis of their characteristics and qualities. Generally, plant taxonomy can be defined as the hierarchical classification of plants on the basis of their morphology and phylogenetic characteristics. Under the hierarchy, the plants are placed in subsets of a superior group. Classifications of plants make use of various elements including their morphology. Plants can be placed into various categories influenced by their structure and shape. The hierarchy of plants follows the basic hierarchy used in the classification of animals with the order being from highest to lowest. The hierarchy is as follows: Kingdom, phylum, class, order, family, genus and finally the species of the plant.

3.2.9: Evolution

The evolution in plants is organic in nature. Various theories have been developed to try and explain the evolution of plants throughout the years. The organic evolution of plants gives the ideology that the recent species of plants have their origins in other pre-existing forms. Comparisons between the plant descendants and ancestors show that there are great differences between the two. The variations are attributed to the modifications that occur in successive generations. Many may attribute the theory of evolution to Charles Darwin though he did not invent the idea as there are those that were before him who had already come up with the notion of evolution. However, Charles Darwin played an important role in providing the much-needed information that would ensure the theory was acceptable. He proposed a mechanism known as natural selection to properly explain why the chances occurred among various organisms including plants. He also highlighted the fact that for such changes to occur, sufficient amounts of time are needed. He stipulated that the first requirement of the evolution process were drawn from uniformity. This was very essential in providing a major link in reasoning. He proposed that there is an existing relationship between geologic time and organic evolution. The relationship is brought about by the randomness in the evolution of various species. This explains the noticeable differences in the physical forms of organisms. (Vyshivkin, D. D. 2017).

Darwin conducted various studies that showcased the existence of an organic evolution. His work was published in 1859 and has provided useful knowledge to both the scientific and non-scientific worlds. This has led to the deviation from the common philosophical and geologic thinkings. Some of the things highlighted in his work include the ideology that plant populations produce progeny at a very significant rate such which enabled them to survive and later on would increase rapidly in number. It also showed that a spectacular progressive increase in population size is less likely to occur. He stated that even though most populations of plants may fluctuate every year, they are highly likely to remain constant over the long term. Darwin also noted that plant populations may struggle for existence as they occur in nature. This is attributed to that fact that each plant individually competes for food and should therefore possess the ability to cope with every facet of the environment. This means that the plant should be able to cope with the physical attributes of the environment including the extreme climates and in the biological aspect, they can deal with the occurrence of predators and diseases. This allows them to continue living and produce progeny.

He also highlighted the fact that the individual plants are different from other plant species. He states that the various plant species possessed similar potential with regard to modification. In his studies of organic evolution, he made a deviation from the results obtained by other researchers. Other researchers had postulated that the modifications of plant structure or genetics are induced by the environment and the newly acquired traits are passed from one generation to another. Darwin suggested otherwise by linking the occurrence of new characteristics to have developed from within the organism by chance. Some of the newly acquired traits are useful to the plant as it allows it to survive while other traits are not entirely useful and in some cases, they do much harm to the plant rather than good.

The evolution of plants is very important as it improves the likelihood of the plants surviving very harsh conditions. In some cases, the new traits allow the organisms to push beyond the challenges presented by the environment. This is the case in plants such as cacti whose leaves have undergone an adaptation process that has undergone an adaptation process allowing them to survive very harsh conditions such as drought. Some plants have been able to successfully pass the new traits to the other generations while others have not. This factor determines the survival or extinction of the plant. The pass info of certain traits to the younger generations of plants is referred to as natural selection. (Charaya, M. U. 2020).

3.3: APPLIED BOTANY

Applied botany makes use of knowledge from pure botany. Applied botany is concerned with the theoretical aspects of botany that give it an application. Applied botany is made up of three subgroups namely botanical pharmaceutical, botanical agriculture and botanical forestry.

3.3.1: Pharmaceutical Botany

Pharmaceutical botany or botanical pharmaceutical is a subgroup of applied botany that pertains to the study of medicinal plants whose applications are in the field of pharmaceutical technology, phytotherapy, and treatment and food supplements. This area of applied botany seeks to identify certain plants on the basis of their morphological, anatomical and taxonomic characteristics, traditional uses of medicinal plants, economic values and flora distribution. It makes use of various information including drug-plant interaction, clinical studies, biologically active components, pharmacological activity and bioactivity for medicinal plants.

There are various areas of research in pharmaceutical botany that includes the identification of harmful, toxic, medicinal and beneficial plants. The study of cytological, anatomical, morphological and taxonomic plants present in naturally distributed flora. It involves studies in the phytogeographical and herbarium database and the ethno-pharmacological uses of medicinal plants. Research is also done on phytotherapy and medicinal plants. Research encompasses analysis of phytochemicals and the variations brought about by the various conditions. It facilitated the generation of information concerning medicinal plants and interactions with human beings and other organisms. For biologists studying pharmaceutical botany, some of the topics in their course include the structure of the cell. Under this topic, they look into the various functions of plants, ergastic and vacuole substances and the cell wall structure. Under meristems, they look into plant permanent tissue where there is dermal tissue characterized by the periderm, lenticel and epiderm. The vascular system is also studied under meristems and looks into the vascular bundles, phloem and xylem. It also looks into the ordinary parenchyma, secretory structures, supporting tissues and reserve parenchyma. Under meristems, the seedling, germ and embryo are looked into. (Vyshivkin, D. D. 2017).

The roots are also studied under pharmaceutical botany. Here the botanist gets information in the secondary growth, primary growth, histology and morphology of plants. The shoot, stem, sprout and bark are studied. It also covers the study of leaves to learn more about their histology, morphology and types. It also looks into the types and morphology of flowers, the types and histology of fruits and the types and histology of seeds. Taxonomy is also a topic under pharmaceutical botany. It looks into several dicotyledons including Asteraceae, Plantaginaceae, Scrophulariaceae, Lamiaceae, Solanaceae, Boragianceae, Valerianaceae, Magnoliaceae and Fagaceae. Under the monocotyledonous families, some of the families studied include Orchiadeceae, Liliaceae, Araceae, Iridaceae and Poaceae.

Pharmaceutical botany is offered in various universities. The course is offered to students with the aim of helping them understand basic microscopical and macroscopical identification methods of medicinal plants used in medicine.

Under pharmaceutical botany, basic information is very essential. Studies on pharmaceutical botany have allowed botanists to identify plants that can be used in treating various diseases. For instance fruits such as beetroot has been found to be very useful in dealing with blood deficiency problems while ginger can help deal with issues of blood pressure. Such studies have been useful in identifying various plant combinations that could be used to make herbal medicine used in treating various health problems. More studies are being done as the side effects of most herbal medicines have not been properly established. This is very useful in ensuring that poisonous plants are not used as herbal medicines.

3.3.2: Agricultural Botany

Agricultural botany is a subdivision of applied botany concerned with plants used in food production, fuel production, turf, fiber and cover crops. Agricultural botany greatly involves research processes as it looks into various factors that could affect plant growth. It looks into plant responses to biotic and abiotic stresses, the breeding of various plants to enhance crop yields in a constrained environment and reproductive biology from the basic level which is the gene to crop production systems. There are various areas of study under agricultural botany including subtropical horticulture, worldwide agriculture and plant breeding among others.

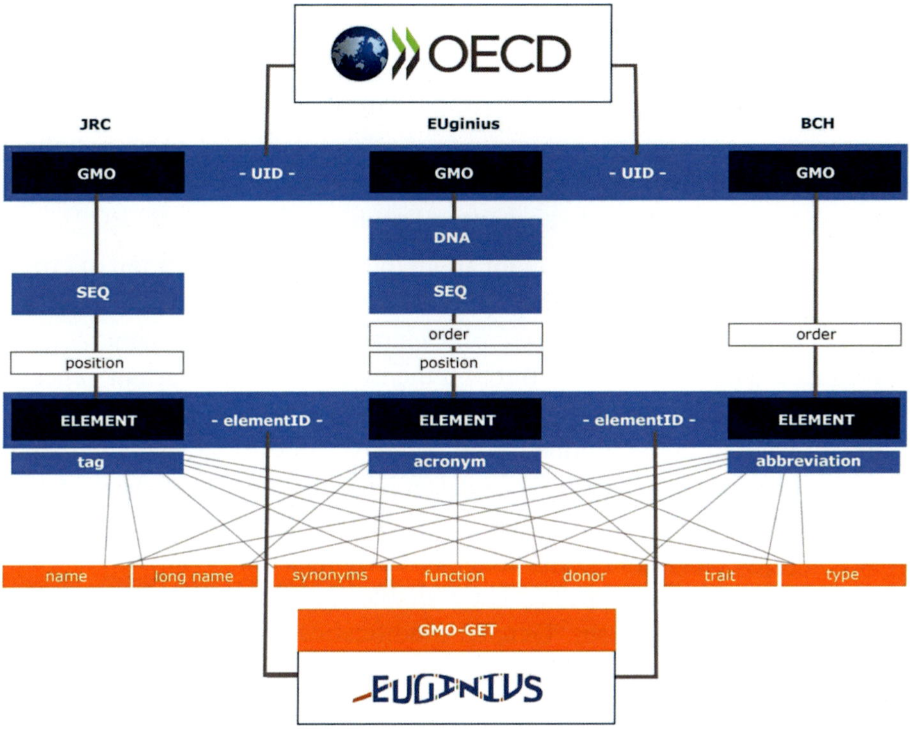

Figure 3.3: Interoperability of GMO-related databases using GMO-GET. The EUginius GMO database, BCHs' LMO & Element Registry and JRC's Central Core DNA Sequence Information System (CCSIS) are able to refer to each other's data via OECD's Unique GMO Identifier (UID) and/or via EUginius' GMO Genetic Element Thesaurus Identifier (GMO-GET elementID). Thus, all three database systems can make use of a harmonized knowledge base identifying explicitly GMOs and their specific components described in terms of their (long) name, synonyms, function, donor, trait and type.

Source: Adamse, P., Dagand, E., Bohmert-Tatarev, K. et al. GMO Genetic Elements Thesaurus (GMO-GET): a controlled vocabulary for the consensus designation of introduced or modified genetic elements in genetically modified organisms. BMC Bioinformatics 22, 48 (2021). https://doi.org/10.1186/s12859-020-03880-0

Agricultural botany gives a clear demonstration of the relationship present between botany and agriculture. There are various plant species that are studied including fruits such as avocado, grapes, peas citrus, tomatoes and olives, cereals such as rice, wheat, maize and wheat and legumes such as cow peas and beans. There are various technologies used in the research process. These technologies are very useful in converting model systems to crops. There are various broad fields under agricultural botany that includes molecular biology and breeding. This may require specialization in certain fields as the scope is very wide. Agricultural botany brought about agricultural biotechnology. This development has been useful in developing the much needed solutions to agricultural problems. (Kartah, B. 2022).

Agricultural botany has brought about the genetic modification of crops. The term GMO's has become more popular over the years as it has been used to deal with problems of food insecurity. Genetically modified manufactured organisms are plants produced through the modification of certain genes in plants. The desired genes for fast growth and disease resistance are made more dominant leading to the production of plants with desirable characteristics. The development of such crops is very useful as it has allowed the production of crops that are more stress tolerant. Stress includes heat, cold and drought. Genetically manufactured crops are also resistant to pests and can therefore survive any incidence of pests. They enable farmers to save more when handling post-harvest losses. Genetic modification of plants has also been useful in controlling the usage of minerals by plants. This is very useful in ensuring the soil does not suffer from issues of infertility. It has also been useful in enhancing the plant's nutritional value. Genetic modification has also been useful in creating tailor-made plants that can be used in various industries such as the pharmaceutical industry. (Charaya, M. U. 2020).

Among the crops that have been genetically modified is cotton. Cotton cannot survive in certain conditions and the occurrence of certain environmental conditions can lead to the destruction of its quality. Bt Cotton is genetically modified cotton. The genetic modification is through the alteration of certain microbes. The newly developed brand is able to survive very harsh conditions. This is one of the ways in which agriculture has made use of botany. In botany, the researcher will learn more about the plant highlighting its weak and strong genetic properties. The development of various technologies allows researchers to suppress certain genes in plants while making others very dominant hence the final product is a plant with desired qualities. Botanical information on the breeding of plants has also been heavily utilized in agriculture to allow the development of plants with desired qualities. The process involves the identification of plants that can be involved in the breeding process. Self-fertilizing plants are also modified such that they become dependent on cross-pollination to allow the generation of new plants that can survive harsh conditions and improve their yields.

Botany has also been used in agriculture in the generation of plants resistant to pests. Certain plants tend to be susceptible to pest infestation. This is attributed to the fact that several nematodes use plants as hosts. Some pests affect certain parts of plants. For instance, the roots of tobacco plants are affected by the presence of a nematode known as *Meloidegyne incognitia*. This leads to a reduction in tobacco yields. Knowledge of RNA interference from botany has been useful in dealing with such a problem. Studies on plants show that the RNA method can be used as a cellular defense method in eukaryotic plants. The method involves a specific mRNA being sliced by a complementary double-stranded RNA. This RNA is involved in the binding and inhibiting of the translation of the mRNA.

The process is considered very advantageous and later on introduced the concept of nematode-specific genes into plants through the use of Agrobacterium vectors. Once the Agrobacterium vectors have introduced the DNA to plants, they cause the production of sense and anti-sense strands in plant cells. As the DNA strands are complementary, they produce the dsRNA while initiating the RNAi. This leads to the silencing of the RNA-specific nematode gene. This means that when the nematode wants to live in the plant, it will not survive hence the plant has become resistant to the parasite. (Vyshivkin, D. D. 2017).

3.3.3: Botanical Forestry

This subdivision of applied botany is a result of the existing relationship between botany and forestry. Botany is very important in forestry. Botanical forestry involves the study of forests to generate information that will allow botanists to better understand them. This is important in allowing its proper use, management and perseverance. Forest botany places much emphasis on forest management and ecology. Forest botany is very important as it allows us to know the species of plants that existed before and have now become extinct or the new species of plants. It also sheds more light on how various Tre species can be protected and how they can be used for various purposes. It also sheds light on the existing relationships between plants and other organisms. In some, the relationship is saprophytic while in others it is parasitic. Botanical forestry is also useful in giving information on the various chemicals that affect the growth in trees and their existence in various environments. It also gives hope to the possibility of generating species of trees now said to be extinct. (Charaya, M. U. 2020).

Chapter

4

CROP PROTECTION IN AGRICULTURE

Contents

4.1 INTRODUCTION

Crops are susceptible to injury, competition, and attack. Air pollution, nematodes, rodents, weeds, insects, and plant diseases are just a few of the many foes that can lower crop yields and prevent a farmer from using some of his farm-stored crops.

For example, insects can quickly destroy a crop. Farmer and scientist attention has been focused on control measures for many years, but full success has not been achieved, and the conflict continues. Control measures may harm honey bees, parasites, and predators that eat insect pests in addition to killing unwanted insects, further complicating the issue.

At least 10,000 insect species are considered to be undesirable. Of these, several hundred species are particularly harmful and need to be controlled to some extent. In addition to destroying food and the forage, pasture, and grain needed to raise livestock, they also spread numerous diseases that affect both plants and animals.(Cheporniuk, H. S. 2015)

4.2: A HISTORICAL REVIEW OF CROP PROTECTION AND ITS SIGNIFICANCE TO AGRICULTURE

The war against pests, diseases, and weeds began when early man transitioned from being a nomadic hunter to a farmer.

Throughout history, we have read about the devastation caused by plant diseases and insect pest outbreaks. In the 1840s, the Late blight of potato epidemic caused by the fungus Phytophthora infestans resulted in a severe famine in Ireland. More than 700,000 Germans perished from famine and disease after World War I due to potato crop failure caused by Late blight and insects.

Since the beginning of agriculture, pests, diseases, and weed competition have decreased crop yield, and farmers have sought to reduce these losses and improve the health of their crops, i.e., they have taken precautions to protect them. In general, the potential for losses due to harmful organisms increases as cultivation intensity increases.

Crop diseases and pests will infect or attack them, and weeds will compete with crops for water and nutrients, resulting in decreased yields. If plants are not maintained in good health, they will not produce consistent, high-quality harvests. In all of this, crop protection products are important. Wheat crop losses are projected to be caused by diseases by 16.7%, insect pests by 11.3%, and weeds by 23.3% in the absence of crop protection and with all other cultivation methods remaining unchanged. (Oerke et al, 2011). The comparative loss rates for weeds, insects, and diseases in Southern Africa are 20%, 30%, and 15%, respectively. Without crop protection, it is estimated that the average global maize production loss would be 59.6% which is 11.7% due to diseases, 19.1% due to insects, and 28.8% due to weeds. In Southern Africa, yield potential is reduced by 10%, 20%, and 40%, for a total loss of 70%. Without crop protection, the estimated global average potato loss due to diseases, insect pests, and weeds is 24.4%, 26.4%, and 22.2%, respectively, for a total of 73.6%. Southern Africa has comparable losses of 40%, 25%, and 15%. Without crop protection, the global loss of cotton is predicted to be 83.5%, which is 10.2% due to diseases, 37.0% due to insect pests, and 36.3% due to weeds. Southern Africa has comparable figures of 10%, 55%, and 25% (Cheporniuk, H. S. 2015).

Crop protection is a broad term for a variety of methods or practices used to safeguard crop yields from various organisms such as pests, weeds, plant diseases, and other organisms that harm crops. Farmers can monitor climate change and catch the emergence of harmful weeds, pests, or diseases early on thanks to cultural protection.

Aside from crops, agricultural fields would also contain weeds, rodents and other small animals, mites, insects, pests, pathogens that cause disease, and birds that would frequently raid the fields. The loss or damage to the crops is primarily caused by all of these factors. Farmers must therefore protect their crops from these pests to produce high yields of crops. So, managing crop protection is crucial before, during, and after cultivation (Khan, S. A. 2020).

4.3: TECHNIQUES FOR PROTECTING CROPS

Farmers can use a variety of crop protection techniques and tools to improve the success of their crops. Insects, rodents, disease, and air pollution can all harm plants and reduce yields for growers. This makes plants particularly vulnerable. Farmers must employ a variety of tactics due to the diversity of hazards. Tillage, crop rotation, the use of natural and man-made pesticides, field monitoring, and other practices are examples of crop protection techniques.

Chemical crop protection

Chemical protection refers to the practice of applying chemicals to crops to protect them from weeds, plant diseases, and insects. These substances are either synthesized or manufactured in chemical factories or plants, like the naturally occurring pyrethrum.

These products are categorized as "Agricultural Remedies" and NOT "Poisons." These include adjuvants, growth regulators, fumigants, bactericides, nematicides, insecticides, fungicides, insecticides, and acaricides (miticides).(Cheporniuk, H. S. 2015)

Chemical crop protection is one of the most economically advantageous methods for protecting plants. Additionally, it facilitates harvesting and ensures year-round growth stability. Here are some widely recognized chemicals.

- Herbicides. Farmers frequently use them to control weeds. Herbicides can be used to kill weeds that are actively growing. Farmers use them before sowing or to repel new pests, but they cannot eradicate weeds at the same time.
- Insecticides. Insecticides are intended to control insects. Some crop protection chemicals are applied to the foliage, whereas others are applied to the soil (e.g., for worm control and against moths and aphids). Insecticides must be applied at the proper time, so farmers scout fields before applying them.

Insecticides are typically effective, affordable, and safe when used properly. However, their benefits may be partially outweighed by their negative effects. Insecticides containing chlorinated hydrocarbons, such as DDT, may leave toxic residues that are harmful to fish, wildlife, and beneficial insects. Residues of these insecticides may be found in meat, milk, and the soil. Some insect species also develop resistance to insecticides composed of carbamates, organic phosphates, and chlorinated hydrocarbons. Only a relentless search for new, safer insecticides and widespread use of non-chemical insect control will be able to overcome these disadvantages (Zadoks, J. C. 2013).

Various forms of organophosphate and carbamate materials are now available. These can prevent the majority of residue-related problems. Malathion and carbaryl, for example, are used to control insects in areas where persistent substances may later be detected in meat and milk. These chemicals may also be used in areas where fish and wildlife are susceptible to harm. These two substances are toxic to a wide variety of insect pests. Contrary to chlorinated hydrocarbons, they can be applied up to a day or so before harvest without harming the majority of crops. However, they are hazardous to the sprayers and must be handled with care.

Some insecticides are effective even in extremely small amounts. This fact has prompted the development of ultralow-volume technology, which enables the dispersion of small quantities of undiluted chemicals with the aid of specialized equipment, offering financial benefits as well as a substantial decrease in the number of chemicals in the environment. Malathion may be effective against grasshoppers, boll weevils, cereal-leaf beetle, mosquitoes, and beet leafhoppers at a rate of six to sixteen ounces (170 to 450 grams) per acre. Granulated chemical formulations offer several advantages for use and application, including a reduction in the required quantity and a decreased risk of adverse effects on beneficial insects and wildlife (Zadoks, J. C. 2013).

Certain insects that prey on cotton, vegetables, and forage crops could be controlled by chemicals ingested by the plant. They are paired with the seed at the time of planting and are referred to as Systemics. When insects attempt to feed on the plant's leaf or stem after it has absorbed the chemical, they die. Beneficial insects that do not feed on plants are not affected.

- Fungicides are used in crop protection to control disease-causing fungi. Throughout the growing season, they facilitate increased yield and preserve stored goods.

To utilize pesticides for crop protection to their fullest potential, it is best to divide the field into zones according to the severity of plant damage. Utilizing only the required amount of chemicals enables the efficient use of resources.

Non-chemical insect control

Cultural management can efficiently minimize pest activity by controlling both mechanical and cultural elements. Changing to new irrigation methods and susceptible plant substitutions is one of the most prevalent remedies (e.g., reducing irrigation can deprive weeds of sufficient moisture and promote root health).

By deploying mechanical crop protection equipment, farmers may physically eliminate pests. Barriers, traps, mulching, and steam sterilization methods are utilized to keep animals and pests out of the field. For many years, scientists have been experimenting with light traps that emit radiation that attracts insects. The tobacco hornworm and codling moth (*Carpocapsa pomonella*) have been combated with a degree of effectiveness.

By utilizing reflective aluminum strips as mulch in agricultural areas to lessen or eliminate aphid assault, cucumbers, squash and watermelons have been protected from mosaic diseases. This technique could replace insecticides, which do not normally kill aphids rapidly enough to prevent agricultural losses caused by the viruses they transmit. (Zadoks, J. C. 2013)

Numerous insects that inhabit storage places can be repelled by heat or cold. Altering the ratios of oxygen, nitrogen, and carbon dioxide in the storage atmosphere can also achieve control.

The reproduction of adult Indian meal moths (*Plodia interpunctella*) was suppressed by 75% when they were exposed to specific sound wavelengths during the egg-laying cycle. Similar effects were produced by sound waves on flour beetles (*Tribolium* species). This method has the potential to be a non-chemical control, albeit additional research is necessary. Other forms of physical energy can kill insects as well. With encouraging results, scientists have explored gamma radiation, high-frequency electric fields, light waves, and high-intensity radio frequencies.

Certain cultural techniques can avoid or lessen insect-caused crop damage. The tasks include crop residue removal, deep plowing, crop rotation, fertilizer application, strip-cropping, irrigation, and scheduled planting. Although these techniques are advantageous, they cannot be depended upon to eradicate all significant infections.

Biological control

The utilization of biological controls has long sparked public interest. Insect pests are attacked by control agents such as parasites, predators, diseases, protozoa, and nematodes. Because nature permits the survival of both useful and harmful insects, biological controls cannot fully replace insecticides. Before a parasite or predator population increases, there must be a significant number of the host species. Occasionally, the pest insect can outnumber the control agents by a significant margin. The Japanese beetle (*Popillia japonica*), the European corn borer (*Pyrausta nubilalis*), the alfalfa aphid (*Therioaphis maculata*), the alfalfa weevil (*Hypera postica*), and several others have been efficiently controlled by parasites and predators.

Microorganisms can be utilized to control parasites. Insects are parasitized by viruses, bacteria, fungi, protozoa, rickettsiae, and nematodes, among others. Numerous diseases are insect-specific and harmless to humans and domesticated animals. It is conceivable for insect pathogens to be manufactured, distributed, and utilized in the same manner as insecticides.

Planting crops kinds that are resistant to pests is the most efficient strategy to control insects. The only issue is that these kinds are not commonly available, and their development takes considerable time. (Khan, S. A.

2020) Using gamma radiation to sterilize male insects before releasing them into the natural insect population is a promising method. In some locations, it has replaced insecticides for the control of fruit flies and screwworms with success. Also promising are chemical attractants that bring insects into touch with trace amounts of pesticide or sterilant.

Taking into account all aspects of pest control, "integrated control," or the coordinated employment of different approaches may be the solution. In California, the spotted alfalfa aphid has been effectively managed by pairing resistant cultivars with a systemic insecticide that leaves parasites and predators unharmed. Before deploying sterile insects, another form of integrated management is the employment of chemical spray in conjunction with bait to decrease a heavy infestation. The tobacco hornworm has been eradicated by the application of pheromone-based attractants in light traps and the management of post-harvest residues. Other examples might be provided, but the primary benefit of these management methods is that they use less insecticide, so helping to maintain the environment. (Cheporniuk, H. S. 2015)

4.4: CROP PROTECTION MANAGEMENT: STRATEGIES AND TECHNIQUES

Methods, devices, and goods for crop protection are widely accessible. In addition, the market's constant evolution gives farmers access to increasingly effective pest control methods. The selection is determined by the agricultural region. The following are the most commonly employed strategies for cultural preservation.

Weed management

Weeds pose a significant threat to plants. They compete for food, water, and space. As a result of their constant growth and dispersion, weeds can suffocate and even kill young plants. To effectively protect crops from weeds, it is essential to quickly identify the threat and comprehend the biological factors at play.

In agriculture, weeds must be controlled because they reduce yields, increase production costs, impede harvesting, and degrade product quality. Additionally, weeds impede the flow of irrigation water, impede the application of pesticides, and harbor pathogens. Earlier methods of weed control included mowing, flooding, cultivating, smothering, burning, and crop rotation. Other techniques, particularly the use of plant-killing chemicals, may be more prevalent today, although they are still useful. A different strategy would be to release insects that only harm the undesirable plant while leaving the crop plants unharmed.

Inadequate biological, mechanical, and cultural control systems have accelerated the use of chemicals since World War II. The use of herbicides has influenced both agricultural cultural and technological practices.

Herbicides come in numerous forms, including wettable powders, granular substances, emulsions, and solutions. Any of these treatments may be applied broadly, in bands, as a spot treatment, or directly to a particular plant part. The chemical is combined with water or oil when creating solutions or emulsions.

Weeding is the process of regulating the growth of weeds. There are numerous weeding methods:

- Herbicides used to control weeds
- Using your hands to pull weeds manually
- Using a trowel and harrow to pull weeds

Before planting the seeds, the field should be cultivated to eliminate weeds.

Spraying is the most common method because dilution permits the uniform application of very small amounts. Instead of granular applicators, spray machines enable sprays to be directed more precisely beneath growing plants, and calibration and rate control are simplified. In contrast, granular formulations have advantages in certain circumstances. Herbicide use must be incorporated into the overall farm program because the optimal application date and rate are dependent on crop stage, weed stage, weather conditions, and other variables.

Herbicides are less expensive when used properly in agricultural production, resulting in a more affordable end product for consumers. Herbicides, for instance, reduce the cost of cotton production by reducing labor requirements for weed control by up to 60 percent. Herbicides have replaced manual labor, which is no longer affordable for farmers in developed nations to use to cultivate crops manually. Equipment for chemical application is widely available (Cheporniuk, H. S. 2015).

When used according to label instructions, herbicides are generally safe not only for the user but also for livestock and wildlife. The greatest difficulty is accidental crop plant damage resulting from drift and soil residues, particularly if residues enter waterways.

Those who produce, sell, and use chemical pesticides and herbicides, as well as those concerned with environmental quality, are debating the future of these products. Chemicals play a crucial role in ensuring a constant and affordable supply of food and fiber. Nevertheless, these substances can be toxic to a variety of organisms and have negative environmental effects. This will necessitate increased vigilance when using chemicals, which may be mandated by law, as well as a greater reliance on nonchemical control methods.

Three factors influence the efficacy of an insecticide application: the selection of an appropriate insecticide, the timing of the application, and the quality of crop coverage. Some insecticides are more effective than others against a particular pest, and others must be applied at specific times. *Bacillus thuringiensis* (Bt) products, for instance, must be applied early in the pest's development, when it is more susceptible to the toxin, to be most effective. (Khan, S. A. 2020)

Pest and insect management

Pests and insects are primarily responsible for crop damage and yield loss. They are capable of destroying the entire crop and consuming a substantial amount of grain. They can reduce crop yield by 30 to 50 percent annually if left unchecked. Integrated pest and insect management is the most effective approach for preventing crop damage. To prevent insects and other pests from damaging crops, pesticides and insecticides are used.

The disease can be influenced by a variety of factors, including plant age, genetics, environmental conditions, and weather, making it one of the most complicated crop protection strategies. As a result, disease diagnosis can be challenging, particularly in the early stages of an infection. Therefore, it is essential to monitor the plants' health regularly and evaluate any early symptoms.

Integrated pest management requires knowledge of the biology and ecology of the pest, as well as the crop production system and agroecosystem. Temperature, for instance, has a substantial effect on the rate of insect development; higher temperatures accelerate development. As a result, the temperature may affect the frequency of insecticide applications. Models based on degree days can be used to predict insect development rates and application timing. In addition to temperature, other factors that influence pest populations include precipitation, host quality, host availability, and the pest's propensity for long-distance dispersal. (Zadoks, J. C. 2013) The pest's typical infestation period and the stage of the crop at which yield loss is most likely to occur will influence management decisions. For instance, if the pest attacks the seedling stage of the crop repeatedly, a preventative strategy may be implemented (e.g., resistant cultivar, insecticide at planting). If the pest attacks the crop infrequently, the decision to eradicate it should be made only when infestations are likely to be economically damaging. Understanding the population dynamics of insect pests in the agroecosystem is essential for making the best management decisions for insect pests on vegetable crops. For instance, a pest may infest a crop (such as alfalfa or wheat) or a non-crop (such as weeds) that does not require control, allowing subsequent generations to develop and potentially spread to nearby vegetable crops and cause damage.

sampling and fiscal limit When deciding whether to use an insecticide or take other measures against an infestation, it is essential to know how much damage or insect infestation a crop can withstand before suffering an unacceptable economic loss. The "action threshold" is the point at which action must be taken to prevent economic loss resulting from infestation or damage. Many vegetable crops have action thresholds that should only be used as a general guide. Adjusting thresholds according to the market, the surrounding environment, the variety, etc. To determine the severity of pest infestations, it is necessary to sample or scout the crop. To determine the size of the pest population, sampling can involve inspecting plants and counting the number of pests or the amount of damage observed, or it can involve setting up traps to catch the pest species. Throughout the growing season, crops are sampled at regular intervals or critical growth stages.

Plant disease management

Another form of crop-damaging factor is pathogenic diseases. Microorganisms such as bacteria, fungi, and viruses exert a variety of effects on plant parts. Utilizing pesticides and biocontrol agents, crops could be protected from these diseases.

Insects are not the only threat to crops. Plant diseases and the microscopic worms known as nematodes have the potential to destroy crops, especially those grown in regions with extreme climatic variations. Plant pests can occasionally restrict the types and varieties of crops that can be cultivated. Sometimes their damage is mistaken for weather-related damage. Epidemics can destroy crops.

Similar to insect control, managing plant diseases and nematodes requires a variety of techniques, including the use of chemicals, the breeding of resistant plants, quarantine, forecasting, warning, cultural practices, and heat treatment, among others. The majority of plant virus diseases are transmitted by insect vectors, so insect control and disease control are linked.

Crop rotation, deep plowing, and the burning of stubble and other post-harvest debris can all aid in nematode and plant disease control. Burning contributes to air pollution and the destruction of organic matter, but it also eliminates above-ground organisms and enables the economical control of chemicals. Another method involves using a propane-gas flame to burn disease spores off of living plants and stubble. The infection of sugarcane is treated with heat treatment in hot-air ovens. Controlling peanut (groundnut) stem rot necessitates plowing under dead plant debris or sowing the seed in a raised bed before applying a pre-emergence weed killer.

For the effective control of an epidemic plant disease, it may be essential to apply chemicals before the onset of the disease. Many governments provide plant disease forecasting and early warning services to farmers. The service is predicated primarily on analyses of temperature, precipitation, humidity, and dew, all of which can promote conditions favorable to disease outbreaks. (Cheporniuk, H. S. 2015)

Use one of the following methods to protect your harvest from disease:

- The use of chemical substances;
- Cultural exchange;
- Extensive tillage
- Establishing quarantine;
- The spread of pathogens resistant to treatment;
- Utilization of heat;
- Conduct routine health checks on the plant.

Plant disease management relies on predicting disease occurrence and targeting disease cycle weak points (i.e., weak links in the infection chain). Through accurate disease diagnosis, the pathogen, the true target of any disease management program, must be identified. Effective disease management requires a comprehensive understanding of the disease cycle, including climatic and other environmental factors that influence the cycle and the cultural requirements of the host plant.

Various techniques, tactics, and strategies for disease management can be categorized according to one or more broad principles of action. It can be challenging to differentiate between these principles. The most fundamental system is based on two guiding principles: therapy and prevention.

The first principle (prevention) applies to any actions taken before infection (i.e., protecting the plant from disease), whereas the second principle (therapy or curative action) applies to any actions taken after infection (i.e., treating the plant). Enforcing quarantines to prevent the introduction of a disease agent (pathogen) into an area where it does not exist exemplifies the first principle.

The second principle is demonstrated by heating or chemically treating vegetative material such as bulbs, corms, and woody cuttings to kill established fungi, bacteria, nematodes, or viruses. Chemotherapy is the application of chemicals to a sick or diseased plant to stop (or eradicate) the infection. Chemotherapy has been utilized extensively, but few attempts have been fruitful. Chemotherapy has been used as a preventative measure against a few diseases of ornamental or other valuable trees, but it must be repeated annually or every few years. For instance, fungicides have been injected into elms to reduce the severity of Dutch elm disease (caused by *Ophiostoma ulmi*), and antibiotics have been injected into palms (lethal yellowing) and pears (pear decline) to reduce the severity of phytoplasma diseases. In all of these instances, the chemotherapeutic agent must be reapplied regularly. In addition, certain "systemic" fungicides, including those that inhibit sterol biosynthesis (SBI) and demethylation (DMI), diffuse into plant tissues and kill newly developed infections.

Crop rotation management

The rotation of crops is one of the most effective and versatile crop protection strategies. However, not all cultures accept the practice equally. For instance, corn can be planted in the same spot for up to five years. Double cropping negatively affects rye, wheat, and sugar beets, but proper crop rotation can increase their yield. This is why farmers occasionally choose a six-field system over a three-field system.

Crop rotation requires a multifaceted strategy. Rotation management requires a balance of field- and farm-level decisions, as well as knowledge of the entire farm and each field. Rotation planning is a fluid, adaptive process on productive farms. The decisions of seasoned farmers are continually weighed against longer-term (short- and long-term) considerations. Cash flow and annual returns must be given precedence when making business decisions. According to knowledgeable cultivators, biological concerns may not always take precedence over market opportunities and logistical requirements. (Khan, S. A. 2020)

Experts implement crop rotations while managing several interrelated factors. Instead of fertilizers and pesticides, organic farmers rely on crop rotations and long-term soil quality to solve problems and ensure field productivity. Their management techniques are constrained by their understanding of biological principles. To satisfy management and economic requirements, a large number of seasoned farmers push and test the biological principles of crop rotation to their absolute limits.

Farmers with expertise plan and execute seasonal, annual, and last-minute opportunistic crop rotations. Their annual plans are founded on distinct objectives. Producing enough profitable crops each year to keep the farm afloat is the greatest challenge. When determining the rotational sequence for each field, farmers must take into account how to rotate labor and machinery throughout the farm operation. Farms and fields must manage crop cultural and harvest characteristics, including the logistics of labor-intensive weeding or multiple harvests, vehicle access, and crop protection from vandals, thieves, and wildlife. When making decisions in the field, farm business considerations, such as meeting market demand and maintaining cash flow, must be taken into account. Expert farmers must constantly adjust their crop rotations and, as a result, their crop mix to the yearly weather conditions.

4.5: IRRIGATION SYSTEM OPTIMIZATION

Soil moisture has a significant impact on plant health and productivity. To determine this, topography, irrigation practices, and precipitation must be considered. Weak cultures and the growth of weeds can be caused by a lack of moisture or an abundance of moisture, respectively. Therefore, farmers must take precautions to safeguard their crops.

Frequently, farmers erroneously equate drip irrigation systems with optimized irrigation. Precision agriculture, however, is much more than that. Innovative irrigation and moisture management techniques can have a substantial effect on the quality of the yield. The world's food production will need to double by 2027 to meet the needs of a growing population. As a result, farmers will have to use more technology to produce more crops more efficiently. In the root zone, nutrients, salinity, microbes, and pests affect plant health. Controlling the amount of water that reaches the plant can be used to manipulate the amount of carbon in the root zone, thereby ensuring consistent yields. With irrigation systems, farmers can use temperature, humidity, evapotranspiration, and real-time monitoring to determine if their plan was successful. With more data, farmers can make long-term plans and remain one step ahead of the competition (Cheporniuk, H. S. 2015).

Working irrigation strategies

Numerous factors determine the type of irrigation system. The first factor is the soil type, as each region's soil has a unique structure and watering needs. The amount of precipitation is the second factor, followed by the size of the field and the proximity of water sources.

Depending on the requirements of the land, the proper type of irrigation can reduce water consumption by up to 30 percent and production costs by a considerable amount.

The majority of farmers can benefit from the three irrigation techniques listed below:

- Sprinkler irrigation
- Watering of topsoil
- Irrigation by drip

Irrigation using sprinklers

Sprinkler irrigation is among the most efficient means of simulating rainfall. It is applicable across numerous industries. Sprinkler or spray irrigation distributes water via a system of pumps, valves, and pipes that are connected to sprinklers.

Because the water in the pipes must be under high pressure, the sprinklers would spray the plants with tiny drops of water. A sophisticated sprinkler system sprinkles plants with water at low pressure and with a gentle mist.

Sprayers are automated and capable of dispersing water in multiple directions. Farmers have them configured for seasonal work. This system is still used on farms around the world, but evaporation and wind can cause up to 35% of the water to evaporate (Khan, S. A, 2020).

Field irrigation

Surface irrigation is one of the oldest types of irrigation techniques. It is a gravity-powered system that uses tons of water. Due to the advantages of drip and sprinkle irrigation, this is currently among the least effective irrigation systems. There are three distinct phases to surface irrigation: advance, storage, and decline.

Water flows through the channel system in the direction of the field during the advanced phase. The second step of the process is to store water around the plant so that it can reach the roots. The final phase of water movement downward and soil drying is the recession.

Flood irrigation is a less efficient method of irrigation, but it is inexpensive and low-tech. This irrigation technique is suitable for farmers who lack access to cutting-edge agricultural technology because it loses less water to evaporation than spray irrigation. This technique is less efficient than drip irrigation because the farmer does not have complete control over water usage.

Drip irrigation

Drip irrigation is one of the most affordable options you can offer your clients. It is highly effective because it can reduce water usage, conserve valuable resources, and deliver water directly to the roots of the plant.

Drip irrigation offers farmers the following advantages:

- Allows farmers to save money and time.
- Efficient utilization of water
- Permits plant rows to remain dry in between rows.
- Prevents weed development
- Protects against illness

Drip line irrigation systems offer the advantage of simple pipe installation throughout the landscape. Due to the pipe pressure, each plant will receive a sufficient amount of water. A programmer can also be used to upgrade the system so that it can be controlled from the comfort of our own homes.

4.6 CROP PROTECTION AFTER CULTIVATION

Now that the crop has reached maturity, it must be harvested. Thus, harvesting involves gathering and cutting mature crops. The immediate post-harvest processes of threshing and winnowing are typically included in the definition of harvesting. Let's investigate both of these processes.

- Threshing refers to the separation of grains from chaff or pods.
- Winnowing: After threshing, the grains and chaff must be

separated. The process of winnowing is the separation of grains.

Before crop harvesting, there are numerous factors to consider. To ensure that crops are not harvested prematurely, they must be thoroughly inspected. Consequently, the seeds are lost and the crop is ruined. If the crops are overripe, their market value may decrease or they may become unusable.

In India, harvesting is typically performed by hand. Sickles are tools used to cut crops. This method is labor-intensive and time-consuming, making it suitable only for small farms. On larger farms, harvesting, threshing, and winnowing are combined using a harvester.

Grain crops must be stored correctly to prevent product loss. The probability of grain loss is significantly greater now than it was before cultivation. Consequently, enhanced protection strategies are necessary. The loss was caused by common pests and rodents as well as environmental factors such as humidity and temperature. Taking a few precautions will prevent this loss. Before storing grains, they should be thoroughly cleaned and dried in the sun, as this helps prevent the growth of mold due to moisture. Fumigation is a method for eradicating insects.

Small investments and straightforward solutions can reduce postharvest losses, at least on the farm level. For instance, the pot-in-pot system is an ingenious, electricity-free food cooling system that could extend the shelf life of perishable foods from days to weeks. This storage infrastructure enables farmers to sell their produce for longer periods and at higher prices, thereby reducing the number of daily trips to the market that many families must make to purchase fresh food.

Hermetic storage bags are an additional method for preserving harvested crops. The bag has an inner and outer layer, and when closed, it becomes airtight, preventing the entry or exit of oxygen and other gases. Cereals such as sorghum, millet, and dried corn can be stored for up to two years in a reusable bag.

To address the various causes of food loss, postharvest solutions have been developed; however, some have not reached farmers or have not produced the desired results. To develop post-harvest solutions that are practical and affordable for farmers, it is essential to inform farmers about successful innovations and include them in the research process. (Cheporniuk, H. S. 2015)

4.7: IMPORTANCE OF CROP PROTECTION

Agricultural practices impact the health of the ecosystem and the quality of food produced. Consequently, agricultural plant protection and the introduction of modern innovations to the industry are essential tools. The technique offers precise mechanisms for maintaining plant health and is continually evolving. The following are the most important advantages of crop protection for farmers:

- Increased yield from smaller areas
- Higher yields that result in more food
- Environmental protection;
- The maintenance of stored goods

Therefore, agricultural plant protection creates all the necessary conditions for a prosperous and sustainable agribusiness.

4.8: CLIMATE CHANGE AND AGRICULTURE

Climate change affects crop protection, which is visible in nearly every aspect of human life, even if we are unaware of it.

The ecosystems of weeds, pests, and pathogenic diseases are affected by climate change. Surprisingly, the alarming rise in global temperatures due to rising carbon dioxide levels affects soil ecology. Researchers and specialists must examine the agricultural implications of these shifting conditions.

It is necessary to examine the effects of global warming on herbivory rates and general insect community patterns. This will determine how pesticides should be utilized to prevent the development of resistance in pest species. One strategy would be to prevent the annihilation of these pests' natural enemies, which could prevent the pest from infesting the crop through natural competition. 98% of the oxygen we breathe and 80% of the food we eat comes from plants, which are vital to life on Earth. However, they are in danger. According to the FAO, plant pests and diseases cause up to forty percent of annual crop losses.

Outbreaks of disease can have catastrophic consequences. The Irish potato famine of the 1840s, which was caused by the fungus late blight, resulted in the deaths of approximately one million people and the forced emigration of another million.

The worst invasion of desert locusts in decades in the Horn of Africa demonstrates that pests can also damage crops. One square kilometer of desert locust swarm consumes as much food per day as 35,000 people, making it one of the most destructive pests in the world. According to the FAO, the outbreak could even cause a humanitarian crisis.

What effect does climate change have on diseases and pests?

In addition to increasing global trade, climate change is one of the factors contributing to the spread of pests and diseases. Climate changes can affect the size, survival rate, and geographic distribution of pest populations, as well as the intensity, development, and geographic distribution of diseases.

According to experts, changes in precipitation and temperature have a significant impact on where and how pests and diseases spread.

According to Tek Sapkota, an agricultural systems and climate change scientist at the International Maize and Wheat Improvement Centre, "in general, an increase in temperature and precipitation levels favours the growth and distribution of most pest species by providing a warm and humid environment and the necessary moisture for their growth.

He explains that when temperatures and precipitation levels are excessive, the growth and reproduction of certain pest species are slowed, and they are eventually eliminated by washing their eggs and larvae off the host plant.

This explains why so many pests migrate from the tropics to temperate regions. Pests favor warmer temperatures up to a certain point. When temperatures are excessively high or low, populations grow more slowly. As temperatures rise, it is anticipated that pest populations will increase more rapidly in temperate regions, which are currently too cold for pests.

Identical patterns can be observed in crop diseases, particularly with fungal pathogens. (Cheporniuk, H. S. 2015)

The Earth's polar orientation

Since 1960, as temperatures have risen, crop pests and diseases have migrated at an average rate of 3 kilometers per year toward the north and south poles of the earth.

In 2015, the fungus known as "tar spot," which is native to Latin America and can reduce maize yields by up to 50 percent, was discovered for the first time in the United States. The disease, which is typically found in tropical climates, has begun to spread to non-tropical regions such as the highlands of Central Mexico and numerous states in the United States.

By 2050, it is anticipated that the southern pine beetle, one of the most destructive insects to invade North America, will have reached the northeastern United States and southeastern Canada.

The Greeks and Romans reported wheat stem rust and sacrificed animals to their gods to prevent the disease from spreading to their wheat harvests.

According to Mandeep Randhawa, CIMMYT wheat breeder and wheat rust pathologist, wheat rusts are adapting to warmer climates and becoming more aggressive. Wheat rusts are one of the greatest threats to global wheat production. As temperatures rise, more spores are produced, which can facilitate the spread of infection and hasten the evolution of pathogenic changes.

According to scientists, stem rust appeared in the United Kingdom for the first time in sixty years. Most likely responsible for the infection-friendly environment, according to the study, are climate changes over the past 25 years.

Elevated CO_2 levels

Insects may be indirectly impacted by rising carbon dioxide (CO_2) levels because they alter the architecture of their host plant's defenses and influence their CO_2 levels.

According to CIMMYT wheat breeder Leonardo Crespo, elevated CO_2 concentrations as a result of human activity and climate change will likely influence pests indirectly through changes in plant chemistry, physiology, and nutritional content.

According to the Intergovernmental Panel on Climate Change, rising temperatures and CO_2 levels could also make pathogens such as fungi more hospitable (IPCC). Although scientists are confident that climate change will increase pests and diseases, it is difficult to predict when and where they will spread. Due to the diversity of pest and pathogen species, climate models can only provide approximations of the potential locations of outbreaks or infections.

Pandemics and pest infestations can be avoided.

In response to these uncertainties, experts are increasingly advocating for a global surveillance system to monitor pest and disease outbreaks and enhance response capabilities. Modern technological innovations, such as the suitcase-sized mobile lab MARPLE, which tests pathogens such as wheat rust in near real-time and provides results in less than 48 hours, make early detection possible. Important tools for alerting policymakers, researchers, and farmers include early warning systems for potential outbreaks.

Breeding pest- and disease-resistant varieties reduce the need for pesticides and fungicides, which is another environmentally friendly solution. CIMMYT works to develop resistant wheat and maize varieties to diseases such as Fusarium Head Blight (FHB), Wheat Rust, Wheat Blast, and Maize Lethal Necrosis (MLN).(Cheporniuk, H. S. 2015)

Chapter

5 GROWING TECHNIQUES TO HELP FARMERS INCREASE CROP YIELD

Contents

With the growing problem of food insecurity, there is an urgent need for farmers to develop all the needed tools to ensure maximum farm production. There are various ways of increasing crop yields discussed as follows:

5.1: SOIL TESTING

Soil testing is described as the chemical removal of elements such as zinc, copper, calcium, phosphorus, sulfur, sodium, magnesium and potassium from soil that are then measured to know the available contents of the soil sample. Soil testing is described as a diagnostic tool that can evaluate the amount of minerals available for plants. Soil testing is among the activities carried out first of farms before the planting process commences. Results from soil tests show the soil pH and the amount of nutrients in soil thereby guiding the farmer if the vest type of fertilizers. It also provides information on the vest types of crops to plant. This makes soil testing very essential as about 60% of crop yields are influenced by soil fertility. This calls for proper soil management. Proper soil management can be achieved by conducting soil tests on a regular basis. Farmers best conduct soil tests when they are fully aware of the importance of the soil tests. (Gemmill-Herren, B. 2017).

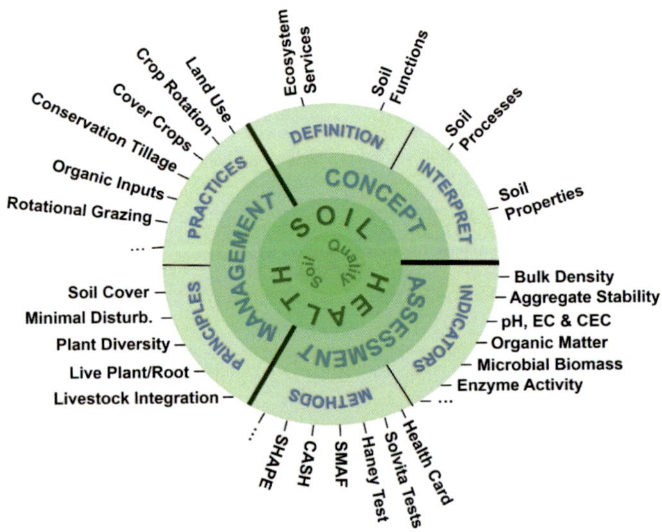

Figure 5.1: Soil testing and soil health.

Source: Guo M. Soil Health Assessment and Management: Recent Development in Science and Practices. *Soil Systems*. 2021; 5(4):61. https://doi.org/10.3390/soilsystems5040061

Some of the objectives of soil tests include it being used in evaluating soil nutrient and fertility as it indicates the amount of nutrients in soil, determination of soil salinity, alkalinity and acidity, providing a basis for determining the amount of fertilizer and manure based on soil value, preventing excessive use of manures and fertilizers, preventing loss of soil fertility, determining the right type of plant to be planted on the given soil and restoration of soil fertility. Frequent soil testing enables the proper selection of the right type of planting program. For instance, when crop rotation is to be practiced, to know the right kinds of crops to plant after certain species of crops have been planted. For instance, if maize has been planted on a given piece of land for some time, it may have to be inter-cropped with beans or other kinds of plants. (Gianessi, L. P. 2013).

Soil testing had become more popular over the years with more tools being developed to facilitate the testing process. More technologies have been incorporated into the testing process so that the device is highly sensitive to produce more accurate results. There are numerous benefits of soil testing both to the farmer and to the soil. Some of the benefits include the fact that soil analysis allows the generation of information on the right kind of fertilizer to be used on the farm, management of soil risks such as degraded lands, soil infertility and soil erosion while ensuring that the farm's profits are maintained at an all-time high. It is also useful in giving more information on the amount of available plant macro-nutrient in soil and the location of these nutrients in soil layers. It also enables farmers to know the nutrients in soil that hinder large farm yields. These nutrients could cause leaching or scorch the plant. (Dietz, K. J. 2019).

It enables the monitoring it soil health by looking into properties such as soil pH which affects the amount and types of nutrients in soil. This generally contributes to the well-being of crops and also increases crop yields. It also influences the type of variable rate application chosen when deciding in the type of crop to plant. Soil testing enables farmers to make better decisions with regard to timely planting, fertilizer application, placement and products to use. Soil testing enables farmers to gain more knowledge on the different types of soil in their land and can therefore make good decisions on the type of crops to plant and fertilizers to apply. It has also proven to be effective in maximizing in-season responsiveness. Soil testing is considered a valuable tool thereby allowing the farmer to know the best way they can improve farm yields while reducing operational costs. It also enables the farmer to better take care of the environment as it is a form of environmental risk management. Soil testing has been useful in improving crop maturity as the farmer will provide all the necessary conditions to ensure high tolerance to pests and diseases. (Gemmill-Herren, B. 2017).

By generating information on the soil's health, the farmer is able to look into the various ways in which they can improve soil health. Soil health is determined by several factors among them the physical, chemical and biological soil properties. Other properties such as soil color, texture and structure go a long way in giving more information on soil health. They however do not give a lot of information on soil chemical composition. The soil and its chemical composition can only be determined through tests. They allow the diagnosis of elements such as the description of soil pH and nutrient level. Such information allows the farmers to decide the best kind of fertilizers they can use to improve soil health. They also know which kinds of nutrients are in soil in high amounts and how best to control such high levels. Healthy soils are the gateway to increased farm yields. These tests also ensure that the farmer does not use a lot of money on fertilizers. In basic essence, it controls fertilizer expenditure. Soil tests show exactly the type of nutrients missing in soil. They will therefore by the right kind of fertilizers ensuring zero wastage. As they have information on the type and quantity of fertilizers needed, they will not waste any money. It also ensures proper management of nutrients such as phosphorus and potassium known to be very limited in the surrounding. As they are a scarce resource, there is a need for proper management of such resources to ensure there is no shortage of such. (Gianessi, L. P. 2013)

In as much as fertilizer application leads to increased farm yields, the excessive application is dangerous to crops. When farmers put fertilizers in soil without proper knowledge of the right nutrients to be applied in the soil. This is dangerous to the environment as some of these fertilizers have the ability to burn the crop. Proper soil testing enables farmers to avoid such mistakes. Excessive use of fertilizers may result in soil and water pollution, harm to aquatic life and leaching of nutrients.

Soil testing also ensures that there is no occurrence of soil degradation. Research shows that about 24 billion tons of fertile soils are lost every year. The main cause is soil erosion brought about by improper soil management. The growing occurrence of land degradation has been said to affect the livelihood of many people and their health as well. Land degradation comes about easily but the reversal process is very costly, takes a lot of time and is difficult. Soil testing is one of the various methods through which soil is well taken care of. It is considered the easiest route to take to ensure that large amounts of money are not spent on solving the issue. By allowing farmers to better take care of soil, soil testing generally contributes to increased food security. As mentioned, soil testing allows all the necessary steps to be put in place thereby ensuring increased farm yields. Soil testing helps farmers reduce pressure placed on soil. This also enables the soil to rejuvenate on its own. Healthy soils also mean that farmers will be able to plant a variety of crops which means that the food supply will also increase meeting the growing need for food. It also means that healthy crops will be produced reducing food supplies that are thrown away.

Soil testing also profits farmers. Production of healthy crops, they are able to generate more profits from their sales while meeting the growing demand for food. Soil testing also enables them to ensure there is a continuous supply of nutrients in their field. They are also able to ensure uniform growth. (Dietz, K. J. 2019). This means that the variety of plants grown is healthier and stronger and can survive any incidence of pests or diseases. It also helps to simplify other activities such as cultivation and spraying. They are also able to ensure uniformity with regard to plant maturity thereby simplifying processes such as harvesting and drying and ensuring quality products are taken to the market. Farmers are also able to find the nutrients needed by plants. They are able to maximize their profits. Other than farmers, soil testing also has numerous benefits to the environment. Proper soil testing ensures that there is the efficient use of plant nutrients through minimized leaching and running off of nutrients into waterways. Production of healthy crops ensures that sufficient amounts of plant residue are left in soil to hold in its place. Plant residue is very essential in the building of soil while reducing water and wind erosion. The outcome of increased farm yields means that farmers do not need to carry out intensive farming practices. (Gemmill-Herren, B. 2017).

5.2 SEED QUALITY

Seeds and other planting materials determine the type of yields the farmers will have. The quality of the planting materials will determine crop productivity. Seed quality is not something to be taken lightly. The type of seeds used will determine crop growing rate and the strength of crops. In crop production, it is more advisable to purchase hybrid seeds sold in agro vet's shop. The type of seed chosen for cultivation will determine whether it will survive the changes in climate such as floods and drought. There are different seeds grown in different times the different crops to be grown in different times. For instance, in dry areas, the drought-resistant maize is grown. Some seed varieties are able to grow in very cool and wet soils. Farmers need to adopt an appropriate seed system. This will enable them to mitigate the effects of the environment. Quality seeds enable farmers to produce higher crop yields that can deal with the challenges presented by the environment. There are several benefits of using quality seeds in the planting process. For instance, it enables the production of bountiful crops (Dietz, K. J. 2019).

Farmers are advised to use good-quality vegetative materials. Some crops such as sweet potatoes cultivated by means of planting materials. Such kinds of materials can be used in various seasons. However, when the same planting material is used continuously, there is a higher likelihood that it will develop viruses. This will result in reduced plant yields. Such kind of practices should be avoided as they will reduce the type of crop yields. Yields in such farms can only be saved when the farmer opts to adopt a better seed system. The kind of seeds chosen should be free from diseases. They should be properly sterilized. This will make them resistant to diseases. Such kinds of seeds are really advantageous as they help farmers improve their yields while ensuring sufficient food supplies. Such success stories were noted in Rwanda where farmers used quality seeds and led to the doubling and tripling of yields. (Gianessi, L. P. 2013)

Quality seeds respond well to various challenges. Throughout the farming process, there are new challenges that come up with regard to food production. Some of them include changes in climate or incidences of pests and diseases. Good quality seeds respond to changes in the environment. They enable farmers to produce the kind of yields that wanted. The good thing is that new varieties of crops are produced meaning new varieties of crops are produced. For the new crop varieties, their seeds are modified over time so that they become more resistant to diseases and pests. This makes them able to handle the new challenges that come about. There are some consumer benefits to using good quality seeds. There are some consumers who will purchase crops based on the types of seeds used to produce the crops. More and more people are becoming conscious of what they are eating. This means that consumers are more likely to purchase crops from farms producing quality seeds. Most of these seeds produce high-quality yields and very healthy crops that will sell very fast in the market. They will be sold for better prices. The increased amount of crop yields means that farmers will be able to earn more. This is very beneficial for small-income farmers. This was well demonstrated by the case in Rwanda. Many farmers invested in the growth of sweet potatoes through the use of quality seeds. The sale of such kinds of products meant that farmers had a lot of products to sell in the markets. This is very beneficial in the development of value chains. This means that small-scale farmers are involved in the production of surplus food amounts and can become suppliers from various manufacturing industries. This has only been made possible through the use of good quality seeds and planting materials (Dietz, K. J. 2019).

Other than the benefit of increased yields, good quality seeds are very useful in ensuring good nutrition among people. Good quality seeds have good effects on the general nutrition of people. Most of these seeds are specially made such that they are rich in certain nutrients namely macronutrients. The specially developed seeds produce more nutritious foods in nature. As most small-scale farmers use food for their own local consumption or sell it to the local market, these nutritious food types will be available for all to consume. Most of the seeds have been useful in producing bio-fortified foods.

The use of good quality seeds has been useful in ensuring that there is a continuous supply of food stuffs. This helps deal with the problems of food insecurity. These seeds as mentioned earlier have more resistance to climate change. Their making is such that its production was due to intensified breeding efforts. Various constraints have been modified so as to achieve the expected results. With the notable decrease and changes in rainfall patterns such that the amount of rainfall is reduced. The drought-resistant seed varieties have become more common. These crops only need water at very few instances. The breeding program is such that there is an exchange of elements such that the plants are fitted with pest-resistant options. When an incidence of such pests occurs, these crops will survive meaning the food harvest will be maintained. These crops are such that despite the fact that they are drought resistant in nature, drought will not affect the quality of the yield. Most of these crops can be post disaster recovered.

The use of good quality seeds has been useful in helping the various agricultural economies deal with natural disasters. They could also help ensure that there are quick recoveries with the occurrence of such disasters. It has been useful in preventing damages that would have been so severe. This is the main reason why most urban populations have become more dependent on such systems. These seeds have been useful in developing networks of suppliers going up to scale. It has enabled the revival of economies. There is a strong correlation between seed quality and farmers' crop yield. Most of these seeds tend to be superior compared to other plant species. They have more physiological purity and are free from diseases. The genetic and environmental factors shed more light on the quality of the seeds used. (Gemmill-Herren, B. 2017).

The challenge however is that in various areas, the agricultural industries tend to overlook seed quality. This has led to very adverse effects including the increasing rates of food insecurity, increased production costs and improved tanking profits. Most governments have not invested in high-quality seeds. This means that there is a need for the domino effect with regard to financial instability. In some areas, farmers are given the mandate to decide on the type of seeds to be used in the planting process. The government works to ensure that these seeds are available to all. Some of the seeds are imported and sold to local farmers while in some countries they have been well modified such that they are able to produce their own certified seeds and distribute them locally and internationally. This means that governments should invest in the production of good-quality seeds. They can support farmers in the development of proper nursery beds to produce good-quality seeds which can be sold locally. They can also look into various ways they can improve the already existing seeds. The use of the genetically engineered crops can also come into play. It enables farmers to eliminate certain traits that they do not want in the crops. This means that there is an open field for more research. Good seeds do not come very easily. They are a result of hard work and sacrifice. The seeds need to be placed into various solutions having different components. Several studies are done to identify the most suitable solutions. They also have to be planted in various soils and conditions to determine where these seeds thrive best.

Some of the good quality seeds are those known to be self-pollinators. Self-pollination is not a good attribute as it could lower the quality of the crop. For this reason, seed varieties developed are unique such that the possibility of self-pollination is completely eliminated. This makes most of the plant breeds to be completely cross-pollinated. Cross-pollination also enables the generation of new plant species. They also attract more pollinators such as bees who are important to the development of food among other vegetables. There are other benefits of high-quality seed crops. (Gianessi, L. P. 2013)

Characteristics of high-quality seed crops

High-quality seed crops are known for their high yields. Most of these seeds are known to have a certain degree of physical purity. This does not affect the genetic purity of the plant. This means that the basis of the seeds is at least

99.5% genetically pure. Plants used in the breeding process should be 100%. This means that the certified hybrid castor is 85%, hybrid cotton is 90% and certified hybrid seeds are 95%. These seeds also tend to be free from various diseases. Certain crop species are usually involved in the process of spreading certain diseases. Some of these diseases are also spread by infected crops. This means that there is a higher likelihood of the diseases being spread among the various groups of plants. These diseases can contaminate the lesions on flowers, pods, leaf tips and leaf margins. There are various kinds of seed borne pests including the Kernel smut of sorghum, the early blight of tomato and the grains smut of pearl millet. These seeds should be sterilized from such diseases reducing the likelihood of such contamination occurring. If the seeds are not maintained at good quality, there is a higher likelihood of them being affected by such diseases. This will greatly affect the crop yields. When the high-quality seeds are used, crop yields will not be affected allowing food demands to be met. (Dietz, K. J. 2019).

Another characteristic of such plants is that they are usually free from weeds. There is a high likelihood of crop yields being affected by the presence of weeds. Weeds compete with plants for nutrients and water. Some weeds are very harmful such that they choke the plant completely. Weeds seeds are very similar to that of normal size crops. This means that it is very hard to distinguish between the two. Most of these weeds will appear harmless to the farmer but are actually very dangerous. Agricultural yields also include those obtained from animals. Weeds can affect such agricultural yields. For instance, if a cow eats the Mexican marigold weed, its milk will be contaminated and therefore cannot be sold to consumers. Farmers need to be careful to ensure that they do not purchase weed seeds instead of the intended crops. This will avoid serious damage. However, some weeds are usually grown for food. This is the case for weeds like pigweeds. There are good quality seeds for such weeds. Planting should be done with high levels of care being taken.

High-quality seeds have high germination and high vigor. Under seed germination, there are terms such as seed germination tests which involve the studying of the ability of seeds to grow at a fast, slow or moderate rate depending on the environmental conditions. Most such tests are performed in the laboratory within a predetermined time. The laboratory allows the controlling of the environmental factors such as moisture, light and temperature. High-quality plants have high germination and have high vigor. The opposite applied to poor quality seeds which record high levels of seed determination. The occurrence of seed deterioration leads to the loss of plant viability thereby causing low germination to become less vigorous. High-quality seeds have high germination and tend to produce high-quality seeds meaning high crop yields. Usually, the low-vigor seeds can be linked to the type of seeds selected.

Another characteristic of such kinds of seeds is that they are usually free from other crop seeds. The weight percentage of high-quality seeds may be different from other types of seeds. This means that the farmer may be vulnerable to allowing unwanted seeds to be planted in their own gardens. This mistake can be easily made as there is a lot of similarity between the various types of seeds. The general appearance of seeds makes it hard to separate seeds mechanically. The differences can be noted after the plants have been grown. The high-quality seeds are at the upper hand as they have a higher likelihood of competing for the needed nutrients. (Gemmill-Herren, B. 2017).

For farmers to ensure that they have bought high-quality seeds, they need to purchase them from reputable seed providers who have years of experience with regard to plant and crop production. They can also give more information on the kind of care that needs to be put in place when handling seeds. All the needed precautions need to be put in place during crop production. Such activities may include isolation, fertilization and harvesting. (Gianessi, L. P. 2013)

5.3: EARLY PLANTING

Early planting can be categorized under the timely planting of crops. This means that seeds are placed in soil at the appropriate time. Early planting has proven to be useful in crop production. Such kind of planting leads to early harvesting of crops thereby reducing the likelihood of damage due to unpredicted weather. Early planting, therefore, increases crop yields. A lot of planning goes into ensuring that plants are early planted. One needs to know what time is considered early planting. This means that date planning is very important. Planting time is usually out of the hands of the farmers as they may not be able to adequately determine the best time for planting. Early planting is identified when all the needed conditions for planting have been achieved. The decision-making process is very important as it contributes to the success or the downfall of the farmer. Some of the advantages of early planting are that it leads to high amounts of moisture in soil that will allow the growth of seeds. Here the farmer does not have to do much with regards to watering the plant as the soil has the needed amounts of moisture. High moisture content also contributes to fast growth in the plant. Early planting therefore eventually contributed to the higher YEILD potential as crops with be grown in optimum conditions. Early planting lowers temperatures for pollination. This means that pollination will take place at a suitable time with higher likelihood of the process being a success. It also positively affects pollination by lengthening the days in which pollination was meant to occur. Early planting also means that plants will be harvested at an earlier time avoiding any damages that could occur due to a sudden change in weather patterns.

The advantages of early planting can be noted from studies in various areas. For instance, in Mississippi, corn production has been greatly affected by late planting. This problem has been noted in most corn production sites. Comparisons between Mississippi and the United States are a demonstration of the importance of early planting. The United States was noted to produce late amounts of crop yields. One of the methods used in determining the most appropriate time for planting is the use of soil temperature. In most cases, it is more appropriate to plant when the soil temperature is maintained at 50-55 degrees Fahrenheit. Early planting has proven to be effective in ensuring that plants do not suffer from environmental stresses. For instance, when dealing with corn, for high growth rates it is important to ensure that there is a large amount of water available more so during silking. In Mississippi, for effective corn growing, farmers should ensure that about 1.5 inches of water is provided to farmers during the process of growing. The amount of rainfall experienced in Mississippi is not sufficient to meet the demand. This means that the amount of water in soil is very important. Early planting enables corn to make use of such benefits. This means that when the soil water capacity is at its maximum, it is the most appropriate time to plant. For instance, if the plants are planted at an opportune time such that pollination occurs in late May and June, there will be a better moisture relationship eventually meaning that flowering of the crops will occur around July and August. (Gianessi, L. P. 2013)

The type of temperatures experienced in such times is very crucial and will affect the yields. High temperatures may not be suitable for the planting process. This is because it may lead to the death of pollen or lead to kernel abortion. When crops are grown when there is stress and high temperatures then pollination will not yield promising results. Under conditions of high temperatures and water stresses, there is an accelerated rate of pollen shedding and the silking process is delayed. This means that the ear of the plant will demonstrate barrenness. It will also cause some stress with regard to grain filing that leads to competition between the reproductive and vegetative components causing a great reduction in crop yields. (Gemmill-Herren, B. 2017).

With early planting, the growth and development of crops are greatly dependent on the temperature at that given time. Early temperature means that the morning soil temperature is about 55 degrees Fahrenheit. This should be present for about three days. This is usually the case from early January to mid-March. During these times the soils are cold, wet and heavy. When the soil has adequate moisture, the seeds can be planted at about one and two inches deep. If the soils are to be irrigated, then the seeds should be planted deeper than 2.5 inches. When corn is irrigated, there may be consistency with regard to the production process. However, there is a risk of such plants being prone to diseases and pests during May. This however is dependent on the type of seeds planted. Farmers need to invest in the right equipment that will enable them to know the most appropriate time for planting. These tools are usually upgraded with the development of new technologies. It is more advisable to make use of the available conditions provided by the environments. All the available evidence is against the late planting of crops.

5.4: CROP ROTATION

Crop rotation is also known as soil rotation. This practice involves the planting of a certain crop in a field at a given duration and after the crop has been harvested, another species of crop is planted in its place. Crop rotation has been useful in improving soil fertility while improving crop yields in farms. Crop rotation has also proven useful in dealing with problems of soil erosion in different areas. Studies show that mono-cropping, planting the same species of crops for longer durations, tends to lead to reduce crop yields. This is because it leads to certain nutrients in soil. This makes the soil loose and unable to properly handle plants. There are different kinds of crops. Some crops use the readily available nutrients in soil while other plants such as legumes fix these nutrients into soil. Crop rotation does not necessarily mean that different crops are planted on the same land at different times. The types of crops planted should be properly determined. When a particular crop leads to the loss of certain nutrients in the soil. The other species should allow the introduction of the lost nutrients. (Van Bruggen, A. H. 2015).

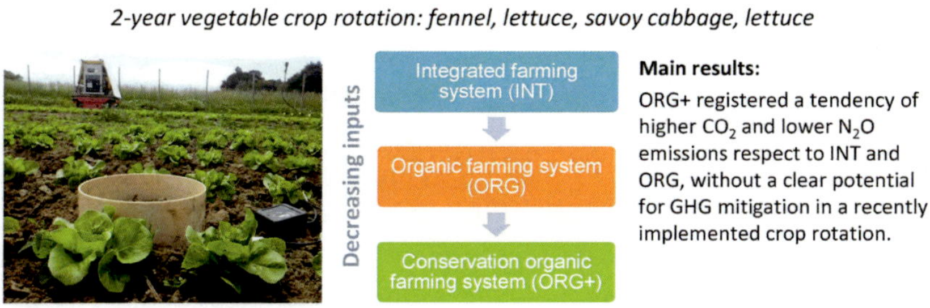

Figure 5.2: Crop rotation.

Source: Bosco S, Volpi I, Antichi D, Ragaglini G, Frasconi C. Greenhouse Gas Emissions from Soil Cultivated with Vegetables in Crop Rotation under Integrated, Organic and Organic Conservation Management in a Mediterranean Environment. *Agronomy*. 2019; 9(8):446. https://doi.org/10.3390/agronomy9080446

When the type of crop rotation carried out is more diverse, the kind of yields and stability of plants is very high. There are various studies that can prove this fact. For instance, in certain unfertilized lands, it was established that higher yields of grains such as corn were generated when it is inter-cropped with soybean. The yields were likened when the soil was fertilized using continuous cropping systems. Higher yields were also noted when corn was inter-cropped with grain sorghum. This was noted in farms that do four-year rotations of these crops. The yields were higher than those in continuously cropped farms. In most of these farms, legumes are inter-cropped with other legumes. The legumes are useful in introducing various nutrients into soil. Intercropping was found to be very useful in ensuring long -term yields while ensuring minimum losses. (Van Bruggen, A. H. 2015).

It was also noted that inter-cropping was similar to the use of the When there are low amounts of nitrogen in soil then addition of N-fertilizer in farm fields has proven to be useful in maintaining the yields of inter-cropped sorghum and corn over a given period of time. Yields were also high when the crops are inter-cropped with crops that can introduce nitrogen into soil. When conducting crop rotation, the sequencing of the process will also affect the type of rotations. (Gemmill-Herren, B. 2017).

Crop rotation has been found to be useful in making soil organic compounds to be at higher amounts compared to Fertilizer-N. The amount of

SOC was found to be higher in soils that experienced more diverse rotations. The amount of nitrogen in the soil can be increased by playing clover crops. Crop rotation enables the building of soil organic matter. It improves soil water retention. It greatly contributes to increased farm yields as it ensures that crops are more resistant to changes in temperature. Crop rotation is not a new endeavor in the field of agriculture. It has been carried out for ages and has over the years proven to be highly effective in improving crop yields as it enables the breaking of weed cycles, reducing damages caused by insects and diseases. Crop rotation has proven to be more useful compared to the application of certain fertilizers. (Gianessi, L. P. 2013)

5.5: INTERCROPPING

This is the practice of growing two different species of crops on the same piece of land at the same time. This practice has been carried out for many years similar to crop rotation. This practice may involve planting a crop that takes certain nutrients away from the soil while planting another which allows or fixes the lost nutrients into soil. For this reason, most grains are inter-cropped with legumes. The legumes will fix the lost nutrients in soil. Intercropping may also occur in the form of planting certain crops alongside some cover crops such as cloves. The cloves fix some nutrients in the soil that will be used by other crops. This practice has proven to be useful in reducing the incidence of pests and diseases. It breaks the cycle of planting a single crop on a given piece of land over long durations of time. This ensures that the crops will not t lead to infestations caused by those plants.

5.6 KNOWLEDGE OF HIGH YIELD POTENTIAL

When growing crops, it is very important that the farmer is knowledgeable about the yield potential of the planting material. Knowing the type of seed to plant, they also know the different types of nutrients needed in soil. For high yields, farmers need to adopt the planting of high-yielding seeds. Such kinds of seeds are known to produce large amounts of crops. The high yield potential is also linked to the use of fertilizers, regular watering of crops and the use of pesticides among other routine practices. In some farms depending on the weather, it may be advisable for the farmer to adopt irrigation on such farms. This will ensure that such seeds are able to access the surplus amounts of water they may need. In most cases, the high-variety seeds are mostly found in soybeans, corn and wheat. They are mostly sold to farmers in the commercial sector as high yields mean that they will generate a lot of profit. Among the effects of using high-yielding seeds is that when the seeds lose some water, only a significant amount of water is needed to ensure it gets back to where it should be. Another effect is that most of these seeds are resistant to disease. Some chemicals are needed to ensure fast growth. The other advantage is that they produce higher yields compared to the regular seeds. The challenge of using such seeds is the fact that they are usually expensive. This means that local farmers cannot purchase such seeds. Some may opt to grow their own seeds. (Gemmill-Herren, B. 2017).

There are several advantages to using high-yielding seed varieties. They include the fact that such seeds tend to have shorter life cycles compared to other types of crops. The short life cycles mean that farmers can grow multiple crops at once. This increases the yield from crops. A good example is with seeds of rice and wheat known to have life cycles of 100 to 120 days respectively. They are different from the other varieties of seed with longer life cycles. This means that farmers are able to save more and get more profit. The high-yielding varieties are greatly dependent in water to ensure that high crop yields are generated. Studies show that with the growing of high-yielding crops, the amount of yield per square meter is very high. Comparisons made between these types of seeds and the regular types of seeds show that the high-yielding seeds need less amounts of water. This means that they save on the amount of water used. When using the high-yielding varieties of crops, minimal amounts of energy are needed. This is the exact opposite when these conditions are not fully satisfied. This means that more jobs will be created to meet the labor demand. The need for large amounts of labor is eliminated when there are sufficient amounts of rainfall. This means that farmers will invest more in other farming operations. (Gianessi, L. P. 2013)

These seed varieties have been said to be scale-neutral. This means that some aspects of the seeds can be well maintained. Such seeds allow farmers to benefit from both small and big farmers will benefit from the growth of such plants. These seeds have been so modified that they can be used by both large-scale and small-scale farmers. The growing of such seeds does not need any special skills. These seeds can he adopted by anybody. They can also be grown in home gardens. The challenge with these seeds is that they are less tolerant to pests meaning farmers will have to place more efforts into ensuring that such crops will survive any incidence of pests and diseases. Large numbers of crops grown using such seeds are usually destroyed by insects. More research is being done to better establish the manner in which these plant species can vest be improved and made resistant to pests and diseases. These varieties of seeds have proven useful in ensuring that there are sufficient amounts of food grown to meet demands of the ever-growing populations. (Van Bruggen, A. H. 2015).

5.7 REGULAR FARM VISITS

Regular farm visits have proven useful in increasing farm yields. This is because farmers will be able to properly assess whether their farms are in good condition. They can check whether the soil needs water or fertilizer application. They can also identify an incidence of pests and diseases in a given field. They can adopt the necessary measures to ensure that high crop yields will be maintained.

5.8 PROPER WATER DRAINAGE

Water is necessary for plant growth. However, if the amount of water entering a given field is not well controlled then it might present a challenge to plant yields. High yields are ensured when plants get the right amount of water. To avoid water logging, different drainage channels are adopted by farmers. There are two main drainage types namely surface and subsurface drainage. The subsurface drainage is established with the aim of managing surface water and ensuring that water run off is well maintained so there is no occurrence of erosion. For this reason, subsurface drainage is defined as the removal of water from the soil profile. In this drainage method, water is allowed to flow towards surface inlets. On the other hand, surface drainage involves the removal of water from the soil surface. Proper water drainage also helps to reduce the incidence of diseases and pests as some diseases thrive in such environments. (Gemmill-Herren, B. 2017).

5.9 FERTILIZER USAGE

We have already established that plants need healthy soils to ensure high yields. Healthy soils can be achieved through fertilizer application as it ensures that soil contains the right amount of nutrients needed to facilitate their growth. Fertilizers are mostly applied either before or after seed planting. Proper care should be taken during the fertilizer application period as the amount of fertilizer applied in the soil should not exceed the needed amounts. Excessive fertilizer application can lead to leaching among other disasters. Fertilizer application is considered central to increased agricultural productivity. It has demonstrated remarkable effects with regard to performance. Therefore, sustainable fertilizer application is needed. Fertilizers are applied depending on the type of nutrients missing in the soil.

For instance, if the soil has a deficiency of nitrogen then fertilizers having nitrogen will be applied in the soil. To determine the type of fertilizer needed, soil tests are done to provide more accurate conditions. (Gianessi, L. P. 2013)

5.10 WEEDING

Weeding is the practice of removing unwanted plants from farms. These plants compete with crops of their limited resources. Weeding is done so that the unwanted plants do not dominate crops. The presence of weeds in soil may seriously damage crop yields. Some weeds contribute to the stunted growth in crops or their death as they take up all the water. Some weeds may host some pests and may Mae crops prone to pests and diseases. Proper management practices should be done when handling weeds so that they do not regenerate shortly afterward. In some areas, there are laws on weed disposal that should be followed. Weeding should be done at the appropriate times and at a given frequency.

Another activity that leads to increased yields is mulching. It is done to reduce the amount of water lost to the environment in the form of water vapor. This practice ensures that plants are able to get the needed amount of water in the soil without stressing. Various technologies can be adopted to make the farming process easier and more effective. (Gemmill-Herren, B. 2017).

Chapter

6

GENETIC AND EVOLUTIONARY ASPECTS OF DIFFERENT PLANTS AND THEIR HEALTH BENEFITS

Contents

6.1: INTRODUCTION

On Earth, there are about 300,000 species of live seed plants. More than 60% of plants have been utilized for medicinal reasons since the Neolithic period. People today harvest therapeutic herbs from wild plants and cultivated plants as a fundamental component of human civilization. Over 10,000 different medicinal plants make up 87% of the Chinese materia medica. Many chemical treatments, including top-selling antimalarial and anticancer therapies, rely on medicinal plants as critical raw materials. Plant extracts and derivatives make up more than one-third of all pharmaceuticals. Because the vast majority of medicinal plants have not been domesticated, there are currently no methods available to improve their medicinal properties and clinical efficacy. Overharvesting concerns the supply of phytomedicine, as evidenced by taxane-producing Taxus plants.

However, successful domestication and augmentation are impossible without a better understanding of the evolutionary history of medicinal plant genomes. Natural selection can be compared to a concentrated and faster version of artificial selection. The discovery and development of plant-based medicines depend on research into the evolution of medicinal plant genomes. This study discusses the benefits, drawbacks, and prospects of the current research topics and a brief synopsis of recent advancements in medicinal plant genome evolution research (Lin, Z. Q. 2017).

6.2: HOW THE GENOME, GENE AND GENOTYPE EVOLVED

6.2.1: Sequencing of the Genome

More progress has been made in genomic research on important agricultural and model plants than on medicinal plants. To understand genome diversity and geodiversity at the molecular level, you must first know how plants came to be, how they grow, how they work, what traits they pass on, and how epigenetics controls them. These basic facts are written into the blueprints of DNA. High-throughput medical plant sequencing has the potential to improve significantly molecular breeding of high-yielding medicinal cultivars and molecular farming of transgenic medicinal strains.

When choosing medicinal plants for whole-genome sequencing research, the following must be considered: Second, representative plants whose pharmaceutical components are precise and have typical secondary metabolism pathways, like Salvia—third, distinguishing plants that are members of a sizeable medicinal genus or family, like *Glycyrrhiza uralensis* (Chinese liquorice; Fabaceae).

Due to a lack of molecular genetic research on most medicinal plants, early genome studies must be done before the whole genome can be sequenced:

- The candidate species should be confirmed using DNA barcoding.
- Metaphase chromosomes should be used to confirm the karyotypes.
- Flow cytometry and pulsed-field gel electrophoresis should be used to check the ploidy level and genome size.

For example, flow cytometry was used to compare the sizes of the genomes of four Panax species, such as by using *Oryza sativa* as a standard. Compared to *P. pseudo ginseng*, *P. vietnamensis*, *and P. stipuleanatus*, *P. ginseng* has the largest genome. A more reliable way to identify species without a reference genome is to use whole-genome shotgun sequencing to do a genome survey. This non-deep sequencing, followed by bioinformatics analysis, is a great way to determine if whole-genome sequencing is a good idea by looking at the genome's size, heterozygosity, repeat sequence, GC content, and other metrics. Also, RAD-Seq, an excellent tool for figuring out how heterozygous a candidate genome is, could be used to make a RAD library and do low-coverage genome sequencing with less representation.

The platform for sequencing the whole genome is chosen based on how much money is available and how well the candidate genomes are evaluated at first. For a short, simple genome, either the GS FLX or the Illumina HiSeq 2500 would work. Still, most plant genomes are complicated, with >50% repeat sequences and >0.5% heterozygosity in their diploid/polyploid genomes. Large insert libraries for sequencing, like BAC, YAC, and Fosmid, can be made, and advanced bioinformatics tools can be used to check the quality of the sequences and put them together. It is possible to link two or more sequencing methods for shotgun sequencing and paired-end sequencing. For example, initial genome assembly using GS FLX and shotgun sequencing can produce 454 contigs that can then be sorted and oriented using paired-end sequencing data from the Illumina HiSeq or SOLiD platforms to make scaffolds. Then, Illumina HiSeq or SOLiD data is used to fill in the gaps between the contigs. The whole genome can be sequenced more efficiently with these methods. (Charlesworth, D. 2010). The genetic and physical maps are essential for putting together complex plant genomes and studying how they work. The genetic linkage map for Bupleurum Chinese was made with 28 ISSR and 44 SSR markers, while the genetic linkage map for *Siraitia grosvenorii* was made with 29 ISSR and 170 SRAP markers. These preliminary results could be used for therapeutic feature selection using markers, map-based cloning, and metabolic gene mapping. Next-generation sequencing and high-throughput array technologies could help with BAC-pool sequencing and high-density genomic mapping. Most medicinal plants don't have clear genetic links between their parents and children. This makes it hard to make broad genetic maps of these plants. (Lin, Z. Q. 2017)

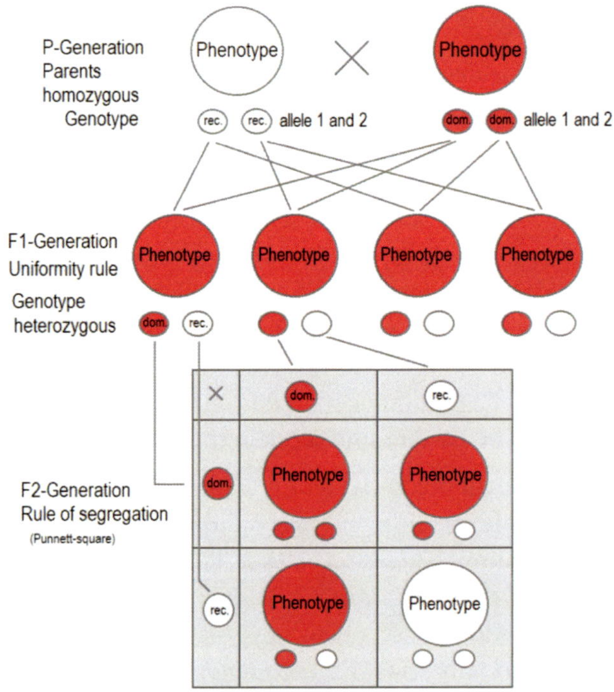

Figure 6.1: Testing of Plant genome-Mendel's law of segregation.

Source: By Sciencia58 - Own work, CC0, https://commons.wikimedia.org/w/index.php?curid=82708882

6.2.2: Genome Evolution of the Chloroplast

Understanding the genetic sequences of chloroplasts (cp), which play a critical role in photosynthesis, is required to correctly grasp most medicinal plants' fast growth, adaptation, and evolution. The mitochondrial (mt) sequence is widely used to infer intraspecific and interspecific evolutionary relationships since it changes much slower than the cp nucleotide sequence.

P. ginseng is regarded as the "crown" plant in Traditional Chinese Medicine (TCM) and is frequently used in clinical treatments and diets to improve health. Next-generation sequencing (NGS) technology reveals the polymorphism and evolutionary history of the P. ginseng cp genome. The cp genomes of the *P. ginseng* cultivars Damaya, Ermaya, and Gaolishen have 156,354 bp, while the cp genome of wild ginseng contains 156,355 bp. These are shorter than the genomes of *Rhazya stricta* cp and 12 Gossypium cp species. The inverted repeat portion has less nucleotide sequence diversity than the big and tiny single-copy parts. The number of genes, GC content, and gene order are identical in DMY and other strains. To investigate cryptic allele variations, high-resolution readings were linked to genomic sequences. The result was the discovery of 208 minor allele locations with minor allele frequencies of 0.05 or below. This revealed that cp genome diversity increased throughout domestication. Polymorphism site density in the DMY cp genome is 0.74 per kilobase, 0.59 in the EMY cp genome, 0.97 in the GLS cp genome, and 1.23 in the YSS cp genome. All minor allele positions in the LSC and IR areas were mapped, and the types of variation discovered at each polymorphism were consistent across all four strains. To adapt to different habitats, purifying selection occurred on minor allele loci of the domesticated cp genome. Focusing on little allele locations within the cp genomes makes it feasible to examine the development of medicinal plant cp genomes and validate a large number of strains and cultivars. (Charlesworth, D. 2010).

Citrus fruits have nutritional and therapeutic advantages that are well-known across the world. Because of the high incidence of cross-pollination between species and cultivars, the number of wild species and evolutionary connections within this genus is unclear. Comparing the Citrus cp genome and creating useful genetic markers will substantially help theoretical and applied research. The Omani lime's whole cp genome was reconstructed using references, and its structure and gene content are comparable with other rosid lineages. Compared to the sweet orange (*C. sinensis*), three intergenic regions and ninety-four simple sequence repeats (SSRs) were efficient in elucidating interspecific relationships. Using these markers might help germplasm preservation and give insight into the citrus domestication process. The evolution of 72 cp genomes from 10 families of key rosid lineages has been investigated in depth. (Lin, Z. Q. 2017)

The Orchidaceae family is well-known among angiosperms for its range of functional and attractive plants. It outperforms both asteroids and rosids in terms of genetic development. Orchid phytometabolites offer a variety of therapeutic benefits, including analgesic, antiangiogenic, and antimycobacterial properties. In its native South Asia, the bulb of the orchid is used to cure various ailments, including respiratory difficulties, gastrointestinal troubles, and skin conditions. The whole cp genomes of *Dendrobium officinale* and *Cypripedium macranthos* were sequenced to understand gene structure and composition better and to look for possible RNA editing sites. Despite similarities in form, gene order, and gene content between the cp genomes of these two species and five other known photosynthetic orchids, the IR/SSC junction and ndh genes in these six species are structured differently. Several Orchidaceae species thrived or perished in the IRs next to the SSC zone. Fifteen protein-coding genes reflect orchids' evolutionary diversity. However, cp phylogenomic research can go beyond the notion that only a few cp markers can detect interspecies relationships.

Bamboo leaves are utilized in traditional Chinese medicine for their anti-inflammatory properties. Medicinal bamboo cupping is an excellent therapy for relieving the pain and stiffness associated with fibromyalgia. Because of its high antioxidant content, bamboo extracts treat infectious illnesses and persistent fever despite the tree's enormous branches and short internodes, analysis of the complete cp genome sequences of 22 temperate bamboos increased resolution along the spinal column of the Arundinarieae group. They provided strong evidence for the great majority of relationships. The cp phylogenomic studies of eight Arundinarieae (herbaceous bamboo) species and 10 Arundinarieae species were confirmed by evaluating their complete cp genome sequences. This study's findings support that the Bambuseae and Olyreae are two different but related groups.

Over 140 species of bulbous perennial plants in the monocot genus Fritillaria have many horticultural and medical applications. Traditional Chinese medicine uses bulb plants of *Fritillaria cirrhosa* as expectorants and antitussives. The phytometabolites in Fritillaria have been demonstrated to have anticancer and cardiovascular effects. Fritillaria species have grabbed scientists' curiosity due to their massive genome sizes, which are all greater than 30 Gb. The phylogenetic reconstruction considered the diversity of the genus' best-known species. The sequencing of three sections of the cp genome resulted in identifying 92 new Liliaceae species and humans. Eleven nuclear low-copy-number genes were investigated in a wide range of species. However, its use in phylogenetic reconstruction was limited. The majority of the currently defined subgenera are monophyletic, according to phylogenetic analysis of a pooled plastid dataset. The most critical TCM species are in the most extensive and diversified subgenus Fritillaria. Plants from Chuan, Hubei, and Anhui provinces all belong to the same subgenus, Bei Mu. The Middle Eastern and Central Asian subgenus Rhinopetalum is a nearby cousin of Japonica, the subgenus with the most known genome size among diploid plants. Differences between genetic phylogeny and morphology-based classifications have occasionally occurred due to phenotypic features' concurrent or convergent evolution. These findings emphasize the significance of information from independently evolved regions, which is especially important considering the enormous nuclear genomes of these plants. Many of the relationships between the significant Fritillaria lineages have been identified, but many more remain unknown. (Lin, Z. Q. 2017)

Many intricate interactions between a plant and its surroundings result in the diverse genetic, taxonomic, and ecological system diversity exhibited in medicinal plants. As a result, environmental complexity and the various ecological processes that contribute to it substantially impact ecosystem diversity. When examining the relationship between climatic/environmental variables and medicinal plant diversity, the repercussions of evolutionary processes must be considered, particularly in places with a significant, uneven species variety. During the Quaternary epoch, southern China's famed "sky islands" were abundant in medicinal plants. Traditional medicine groups, as well as the plant genera Pedicularis, Clematis, Aconitum, and Delphinium, are quickly increasing and diversifying. As the super-barcode of the organelle scale, the cp genome sequence may be utilized to analyze phylogeographic trends and intraspecific variation within the same species across different locales. Sequencing the cp genome and applying the results to populations can provide information on the frequency and amount of intraspecific differentiation. Phylogenetic reconstruction gets more complex when cp data is utilized to infer the interpopulation connection. (Charlesworth, D. 2010).

6.2.3: Genome Evolution of the Mitochondria

The inquiry of the first known incidence of mitochondrial translocation (MT) gives significant evidence for various major evolutionary hypotheses, such as lateral gene transmission. Although the nucleotide substitution rate in the mt genome is one-tenth that of animals, the gene order in the mt genome may develop swiftly in terrestrial plants. As a result, the cp genome sequence may be utilized more precisely than the mt genome sequence to determine the evolutionary relationship between medicinal species. The genome sequence analysis can give insight into the mt genome's history. Terpene synthase was also present in mt, implying a role in secondary metabolism. (Stadler, P. F. 2013)

Rhazya stricta, a plant of the Apocynaceae family, has long been used medicinally in South Asian and Middle Eastern societies.

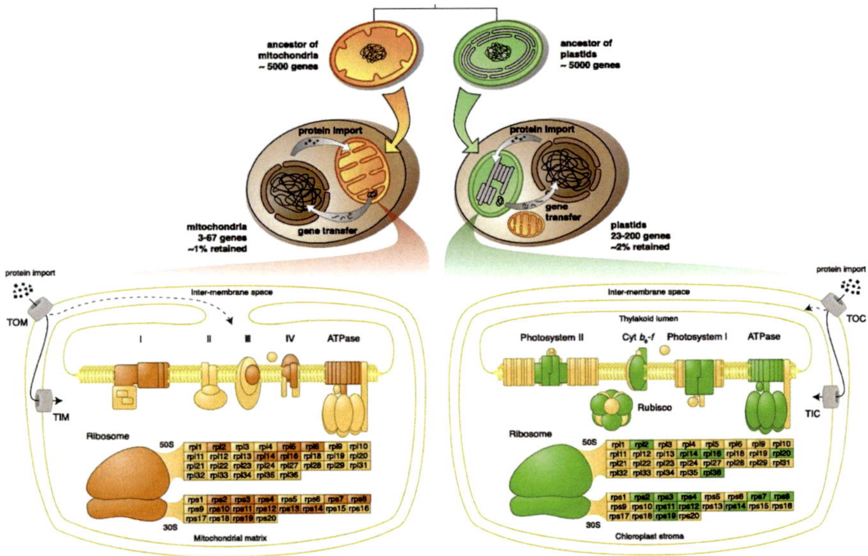

Figure 6.2: Parallel evolution of mitochondria and plastids. The ancestors of both organelles were prokaryotes with nonreduced genomes encoding around 5,000 genes. During the course of endoymbiosis, genes were transferred from organellar host nuclear genomes, and the corresponding gene products were imported back to the organelles. The initial genome size of several thousand dwindled to 3–67 in mitochondria and 23–200 in plastids. The lower part of the figure shows the parallels between the retained genes in mitochondria and plastids (oxidative phosphorylation, photosynthesis, and ribosomes). Organellar-encoded genes are colored brown for mitochondria and green for plastids. Schemes for oxidative phosphorylation and photosynthesis were adapted from Allen (2003).

Source: Maier UG, Zauner S, Woehle C, Bolte K, Hempel F, Allen JF, Martin WF. Massively convergent evolution for ribosomal protein gene content in plastid and mitochondrial genomes. Genome Biol Evol. 2013;5(12):2318-29. doi: 10.1093/gbe/evt181. PMID: 24259312; PMCID: PMC3879969. [accessed 15 Nov, 2022]

Evolutionary histories and intercompartmental genome transfers of eight asterid mitochondrial genomes are revealed in the Rhazya cp, mt, and nuclear transcriptomes. The gene structure and content of Rhazya are identical to those of angiosperms. The recombination-induced repetitions transferred DNA from the cp and nr genomes, and bidirectional DNA transfers between mitochondria and the nucleus comprise the 548,608 bp mt genome. Transit peptides have been employed to transport mt genes to the nucleus, such as sdh3 and rps14. Although there are two copies of Rps14 in the nucleus, only one is active and contains the mt-targeting transit peptide. According to evolutionary evidence, this gene in Rhazya experienced a single nuclear insertion event before being replicated. The very varied sequence of Sdh3 and Rps14 and the evolutionary dispersion of gene losses indicate that they were likely transmitted across asteroids multiple times. The genomes of eight asterid mts show a convoluted evolutionary history for this resistant eudicot group. The cp and nr genomes differ significantly in terms of repetitions, genes, introns, and the quantity of foreign DNA. Genomic data enables an in-depth examination of genetic inheritance patterns. (Lin, Z. Q. 2017)

6.2.4: Genome Evolution of the Nuclear Genome

The entire cp genome must be analyzed to understand the evolutionary relationship between fast-developing species like the Zingiberales. The cp genome cannot be used to calculate population increase since it is similar to a single gene locus and reflects the results of a single coalescent random process. The nr genome contains most of a medical plant's genetic history.

High-throughput sequencing and the statistical analysis that go with it will substantially help future breeding efforts by offering new insights into nuclear genome and transcriptome development. Coix, a Poaceae genus with several species of varying ploidy levels, is closely related to the Sorghum and Zea genera. Traditional medicine practitioners in East and Southeast Asia frequently prescribe using *C. lacrymajobi* pills. *C. aquatica* is a native Chinese species with four cytotypes, three of which successfully reproduce and one of which does not. The majority (76%) and majority (73%) of the HG genomes of *C. lacrymajobi* and *C. aquatica* are made up of LTR retrotransposable elements. Genomic analysis Evidence from many copies of the same sequence, on the other hand, shows that the two species are substantially distinct in evolution. CentX is one of three satellites in the *C. aquatica* HG constellation. Precision karyotyping and fluorescence in situ hybridization revealed that *C. lacrymajobi* is a diploidised paleotetraploid species, whereas *C. aquatica* HG is the product of more recent hybridization. The occurrence of coexisting repetition groups and sequence similarities indicates that these Coix species are more closely related to Sorghum than Zea. (Hackett, J. D. 2011).

Plant families and genera sequenced their whole genomes include *Capsicum annum*, *Coffea canephora*, *Brassica napus*, and *Phalaenopsis equestris*. Scientists are comparing the genetic sequences of the domesticated pepper Zunla-1 and its wild ancestor Chiltepin to learn more about the processes of domestication and specialization within the Capsicum genus (*Capsicum annuum var. glabriusculum*). The pepper genome comprises around 81% of repetitive sequences and 34,476 protein-coding genes. However, due to the activity of retrotransposons, these components are no longer involved in genome synthesis. We compared the genomes of domesticated and wild peppers using 20 resequencing accessions to find the genetic signature of artificial selection and build a list of potential domestication genes. The existence of tandemly duplicated genes that operate as a dosage-compensating buffer might explain the wide range in pepper heat. Using data from the tomato, potato, and capsicum reference genomes is critical to generate new Solanaceae species, such as the well-known Atropa medicinal plants.

The successful sequencing and assembly of the complicated heterozygous genome were among the most amazing undertakings. The decoded heterozygous genome of *C. canephora* demonstrates that the chromosomal gene order of asteroid angiosperms has not changed. Despite the lack of whole-genome duplication, the genomes of Solanaceae species include various gene family expansions that are unique to those organisms. Examples include anti-aphid genes, an alkaloid gene, a flavonoid enzyme involved in secondary metabolite formation, and a neosynthetic methyltransferase (NMT) gene. Caffeine's effects on NMTs differ from those of other drinks, such as tea and chocolate. Caffeine has a polyphyletic background and convergent evolution in its synthesis since the history of caffeine biosynthesis in eudicots changed between species due to multiple tandem duplications. By sequencing the heterozygous genome of the tropical epiphytic orchid *P. equestris*, researchers gained a deeper understanding of the peculiar crassulacean acid metabolism (CAM).

Overall, the completed genome comprises 29,431 putative protein-coding genes and genes that may participate in self-incompatibility pathways, improving fitness and survival by assuring genetic diversity. The development of CAM photosynthesis by P. equestrian was aided by a combination of gene duplication and a second paleopolyploidy event before the widespread dispersal of orchid clades. The distinct appearance of orchid blossoms may be attributed to genes from the MADS-box C/D, AP3 B, and AGL6 families. Long terminal repeats (LTRs) are the most common type of TE, followed by TEs with lengthy interspersed elements (LIEs) (Charlesworth, D. 2010).

The sequencing of the genomes of more than 40 plant species may have implications for agriculture, healthcare, renewable resources, and sustainability. More study is required to completely comprehend the dynamics of gene families indicated by DNA sequence homology. In a comparative genomics scenario, gene synteny, sequence homology, and protein-based hidden Markov modeling may be employed to identify NB-LRR (nucleotide-binding leucine-rich repeat) type multidomain resistance (R)-proteins implicated in plant innate immunity (HMM). When researchers examined the intraspecific and interspecific diversity of R-proteins in the genomes of 12 eudicot plants, they discovered 2,363 NB-LRR genes. Approximately 50% of R-proteins are duplicated, and early polyploidization processes produce copies of 22% of genes (ohnologs, whole-genome duplication duplicates). Positive Darwinian selection operates on all forms of R-gene duplications, including tandem (mean = 1.59), ohnolog (1.36), and singleton (1.22) duplications. We discovered orthologous synteny between one of four sets of NB-LRR "gatekeeper" loci in all of the genomes studied. If these loci are susceptible to genetic modification, they might be helpful in plant breeding. The large diversity of multidomain R-protein clusters identified across the eudicot order may reveal evolutionary processes favoring the diversification of the plant's innate immune system. More plant species must be sequenced to determine if or not more functional NB-LRR genes exist.

The existence of 450 000 plants at any given time demonstrates the natural environment's immense potential. As a result of periodic genome duplications in plants, an extensive range of helpful medicinal chemicals have arisen, the vast majority of which are engaged in plant defense and disease resistance. Furthermore, they provide an enormous chemical area for developing innovative medications. Terpenoids, benzoxazinoids, steroidal glycoalkaloids, and glucosinolates are produced by multigene secondary metabolic pathways that most likely evolved via gene duplication. The rapid emergence of complex characteristics, such as the almost 200,000 secondary metabolites seen in angiosperms, is assumed to be due to genome duplication.

The majority of angiosperm genomes contain only one copy of each gene. Single-copy genes from various species may be discovered, and their evolution may be approximated using the currently available set of 29 high-quality sequenced genomes. The number of single-copy genes was inversely related to the number of duplicate blocks. The bulk of single-copy genes in the body (17% of the total) are present in organelles and perform crucial roles in binding and catalysis. In genomes with a high atomic number, one-copy genes prevail. Single-copy genes in eudicots are more prone to codon use biases than genes in many copies. (Lin, Z. Q. 2017)

RNA-Seq data indicated a disproportionately high frequency of single-copy gene expression. Grass genome copy number (GC3) falls as Nc rises, in contrast to most other species (sufficient number of codons). According to the Ka and Ks values, single-copy genes are more conserved than multi-copy genes. High-copy family genes are more vulnerable to selection effects on alternative splicing than low-copy family genes. Using many copies of single-copy genes, a high-resolution phylogram was created. (Hackett, J. D. 2011).

6.3: IMPORTANCE OF DIVERSITY IN GENETIC ASSESSMENT IN CROP PLANTS AND ITS RECENT ADVANCES

Because of the abundance of available plant genetic resources, plant breeders may generate new and improved cultivars with features desired by farmers and other breeders. Humans have relied on the natural genetic variability of crop species to meet their food demands since the dawn of agriculture. The current objective is to increase global food production to fulfill the requirements of an expanding human population. The advent of high-yielding and fertilizer-responsive dwarf hybrids and cultivars, mainly wheat and rice, in developing nations such as India sparked the green revolution. This problem persists due to various factors, including long-term practices such as extensive single-gene cultivars, genetic erosion, and the loss of essential and adaptable genes.

While science, technology, and agriculture have advanced significantly, whether there will be enough food to feed everyone on the planet by 2050 remains unresolved. Because the world's population is predicted to reach over 9 billion in 2014, this problem was recently addressed during the World Food Prize presentation. Despite this, no therapeutic alternative has been discovered. One option for reducing the consequences of climate change is to lower the average amount of food and water available to each human. Agriculture must be safeguarded to ensure its continuing position as the primary source of human subsistence and a substantial economic contribution. Maintaining a museum as a record of the numerous spiritual and cultural traditions of civilized people worldwide is critical. Agriculture also requires a supply of growing and cultivable crop species. While increasing food production can significantly influence environmental conditions through the propagation of reproductive and adaptable genes, long-term increases in food production are associated with worsening ecological conditions. The need to store genetic material will be underlined, as will the present level of post-genomic analytic techniques. Plant and animal breeders have altered genetic norms by introducing beneficial genes and eliminating negative ones. Because of recent technological improvements, the number of breeding cycles necessary for genetic change has been lowered, and breeding accuracy has risen. (Lin, Z. Q. 2017)

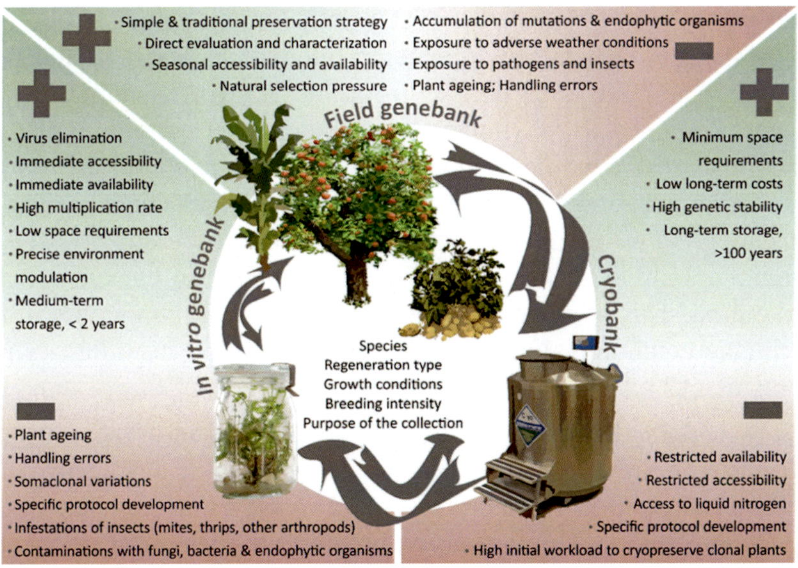

Figure 6.3: Pros and cons of storing crop genetic resources in field genebanks, *in vitro* and through cryopreservation. Arrows indicate the potential source material and target approach to maintain plant genetic resources for food and

agriculture (PGRFA) and can be specific for each plant species. Source: Panis B, Nagel M, Van den houwe I. Challenges and Prospects for the Conservation of Crop Genetic Resources in Field Genebanks, in *In vitro* Collections and/or in Liquid Nitrogen. *Plants*. 2020; 9(12):1634. https://doi.org/10.3390/plants9121634

The combined effects of an increasing human population, urbanization of agricultural regions, and industrialization are wreaking havoc on the planet's biodiversity. Environmental stress, land degradation, deforestation, urbanization, and coastal development threaten plant species, particularly food crops. The Irish potato famine and the Southern Maine leaf rot outbreak in the United States are two examples of food crises induced by extensive cultivation of genetically identical crops. Inadequate knowledge of PGR hampered its application to cultivated crops during the green revolution, putting landraces at risk. The green revolution's purpose was to increase farmers' earnings by assisting them in developing new crop types with greater yields. The Consultative Group for International Agricultural Research established gene banks and domestication research organizations to ensure that PGR is preserved in the world's most essential, stable food crops. Corn, wheat, barley, rice, and potatoes have domestication centers.

According to the Rio de Janeiro Earth Summit, biodiversity is the single most significant factor influencing interactions between and within ecosystems. The main issue derives from the fact that contemporary crop types were designed to maximize yields under ideal conditions. Because of changing stress conditions, these cultivars are inappropriate for low-income farmers in moderately productive areas. Low-input agricultural systems are great for maintaining and cultivating rare and historic cultivars and landraces, and these farms may be more resilient to unexpected drops in food output. This is especially true in sections of Ethiopia with difficult agro-climatic conditions, limited technical growth, and insufficient market mechanisms. (Charlesworth, D. 2010).

Is it essential to lose all of our genetic diversity? Conservation genetics' significant aims are to preserve genetic variety at various scales and to develop tools for population monitoring and analysis that may be utilized in conservation planning. Human survival requires a diverse gene pool. Even though genetic diversity can only be evaluated for populations or species, studies and conservation efforts rarely focus on a single member of that population or species. One technique for disentangling phenotypic and genetic variation is to focus on quantitative features or attributes that fall into specific categories and are impacted by a small number of relevant genes. Because changes in enzyme activity may be used as a proxy for genetic variation, protein electrophoresis is a promising approach for discovering and analyzing genetic variety. Genetic variations can also be studied by studying DNA nucleotide sequences.

In cross-pollinated plant populations, inbreeding is common. Fitness deficiencies caused by unfavorable gene recombination are more prevalent in small outcross groups. The genetic bottleneck increases the likelihood of agricultural extinction due to significant population losses, loss of genetic diversity, and increased sensitivity to infectious diseases and pests. This has ramifications for the entire wild animal population. Genetic models employ the proportion of original heterozygozygosity retained in each generation to compute the optimum population size, often less than the actual population size. This results in a 5% decline in heterozygosity in the population each year. As a result, severe bottlenecks limit heterozygozity and genetic diversity in a population. If plant breeders wish to continue employing a particular trait in their crops, they must keep their people large enough. Researchers must first choose an appropriate population size and representative samples from that group to get reliable estimates of genetic variation (Lin, Z. Q. 2017).

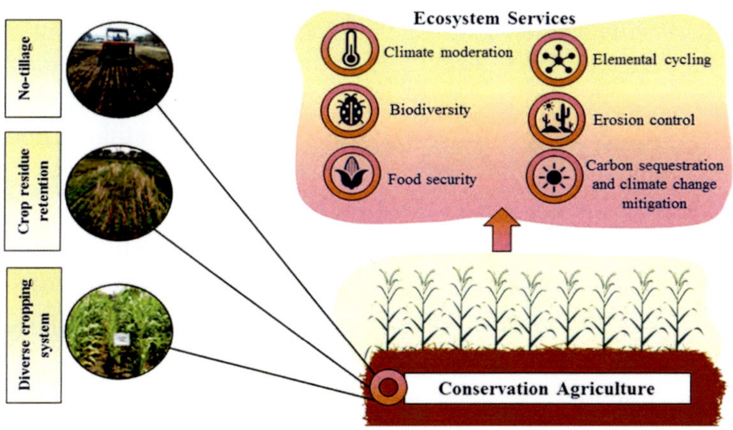

Figure 6.4: Conservation Agriculture.

Source: Jayaraman S, Dang YP, Naorem A, Page KL, Dalal RC. Conservation Agriculture as a System to Enhance Ecosystem Services. *Agriculture*. 2021; 11(8):718. https://doi.org/10.3390/agriculture11080718

Some of the most severe and immediate consequences of climate change on agriculture and food supply are projected within the next decade to two. The present output level will impact how much climate change affects a civilization. Climate change is expected to exacerbate our already severe pollution concerns.

Food production systems may be more vulnerable to pest and disease outbreaks due to climate change due to their reliance on specific cultivars cultivated under ideal circumstances. Reduced annual food production will raise pressure to utilize unsustainable practices, which will have a disastrous effect on biodiversity in and around agricultural areas. In truth, most developing economies have previously encountered similar issues. As a result, adaptations to climatic extremes and other phenotypic changes use less of the genome. The genetic diversity of a population or species is critical to its capacity to adapt to new situations. If this scenario continues, no species will be able to survive in the wild. However, the genetic mix of a crop species determines how well individual members of the species adapt to their biotic and abiotic surroundings.

People may potentially disperse around the biosphere once several generations have passed. As the local environment deteriorates, a population with lesser genetic variety and more uniform features may go extinct. If these populations disappear, the viability of the plant species is jeopardized. For example, the Florida yew is a severely endangered conifer in the Americas. The species flourished in the early Holocene when the southeast of North America saw milder temperatures and more precipitation than it does today. *T. taxifolia* remained in the south when Holocene temperatures climbed. To what extent is this not self-evident? It has only been spotted in Florida's panhandle and southern Georgia along the Apalachicola River. Plant species that are physiologically stressed, such as *T. taxifolia*, are more likely to encounter extinction vortexes, downward spirals of tiny populations, and other similar occurrences.

The extinction cycle reaches a tipping point when genetic variation is eliminated. There will be a growing requirement for agricultural adaptation to climate change. Hence gene banks must be established. Plants housed in gene banks need to be more diversified. Therefore, they must collect what is now referred to as "minor crops" and analyze them for various attributes. There might be severe consequences for retail internet purchases. (Charlesworth, D. 2010).

6.4: GENETIC DIVERSITY IN CROP PLANTS: ITS ASSESSMENT

In the postgenomic era, DNA marker analysis, mainly single nucleotide polymorphism, and morphological and biochemical characterization/ evaluation in the pregenomic period are used to assess genetic diversity in plants both within and between populations. Markers may exhibit inheritance patterns similar to those found in other characteristics. The quality is codominant if the genetic makeup of homozygotes and heterozygotes can be discriminated. Codominant markers can provide more information than dominant markers.

Even though they are based on easily observable characteristics such as flower color, seed shape, growth habits, and pigmentation, morphological markers frequently necessitate large tracts of land for field experiments, which may make them more expensive than molecular analysis in Western countries and similarly costly in Asia and the Middle East due to labor costs and the availability of these resources. These marker phenotypes' phenotypic plasticity allows for evaluating evolution in the context of genotypic and environmental variety. For genetic interference to be successful on adult plants in the wild, seeds with spines, panicles with bristles, and a variety of flower and leaf hues are still required.

The second form of genetic marker is biochemical markers, sometimes known as isozyme allelic variations. Specialized staining and electrophoresis are used to locate them. Several isozyme signals have been discovered in nature. They have a basic inheritance system and can detect alterations at the functional gene level. Plant material of any size can be utilized for testing. Enzyme markers are few, restricting our ability to investigate the entire breadth of genetic variability. Despite their intricate construction and particular problems, these enzymes are not unique. (Lin, Z. Q. 2017)

The third and most prevalent sort of genetic marker is a molecular marker. They are used in genetic and molecular variation studies to identify various DNA molecular markers. These markers identify chromosomal insertion, inversion, duplication, and deletion disorders. Because these markers are located near or connected with the genes that govern the target attributes, they do not affect the phenotypic expression of those features. These markers show codominant as well as dominant inheritance patterns. A battery of markers can reveal many genetic characteristics (they can be dominant or codominant, amplify anonymous or characterized loci, contain expressed or non-expressed sequences, etc.). A molecular marker is a piece of DNA (or RNA) that can be precisely found with a probe or a customized primer, allowing unequivocal identification of the chromosomal characteristic it represents and its 3′ and 5′ surrounding areas. Molecular markers can be connected to how a feature shows itself phenotypically or not. They provide considerable advantages over standard phenotype-based investigations since they are stable and observable in all tissues, independent of the cell's growth, differentiation, development, or defensive attitude. They are also shielded from the impacts of pleiotropy, epistasis, and the environment. We won't delve into pregenomic approaches since we're more interested in the genomic discoveries and how they might be utilized to assess agricultural genetic diversity.

Suppose taxonomy categories based on visual evidence are to be interpreted as suggestive of patterns of genomic differentiation. In that case, a detailed study of germplasm's observed molecular genetic diversity is required. Understanding population structure, allelic diversity, and germplasm diversity may aid breeders if it allows them to make more efficient use of genetic resources for cultivar development with fewer breeding operations. Because molecular markers may give precise and timely information, their application in characterizing germplasm has grown in popularity. Each DNA marker's abbreviation may be found in the Abbreviations section.

Because DNA markers are widely available, they are often used. Mistakes in tandem DNA replication, structural rearrangements, and substitution mutations are some of the DNA faults that contribute to their creation. Natural selection has little effect on these markers since they are often situated in non-coding regions of DNA on chromosomes. In contrast to other features, DNA markers are infinitely numerous and unaffected by environmental factors or the plant's developmental stage. Parental selection, cultivar identity and purity evaluation, heterosis investigation, finding genomic regions under selection, marker-assisted backcrossing, and marker-assisted pyramiding are all applications of DNA markers in plant breeding. (Charlesworth, D. 2010).

The methods used to uncover molecular markers are divided into hybridization, polymerase chain reaction (PCR), and DNA sequence analysis. Restriction fragment length polymorphisms (RFLPs) are a form of hybridization-based marker that was first created for use in human genetic investigations in the 1980s.

When two or more individuals of the same species have their genomic DNA digested and hybridized with a set of markers, the sizes of the resultant

DNA fragments may be compared using a restriction fragment length polymorphism (RFLP). The genomes of the essential cereal species may be compared and contrasted using RFLPs. Among the species, the species are Rye, wheat, maize, sorghum, barley, and rice. RFLPs have advantages such as codominance, robustness, reliability, and data that may be used for a wide range of populations. RFLPs have several disadvantages, including poor polymorphism in distantly related lines, significant DNA needs, a high labor effort, and a high cost. Today's polymerase chain reaction-based marker technologies are more accurate and use less plant material for DNA extraction. Several copies of polymorphic DNA strands were synthesized in a polymerase chain reaction (PCR) machine using genomic DNA and randomly chosen forward and reverse primers. The first PCR-based markers were created using these primers. RAPDs offer several advantages, including their speed, ease of use, low cost, ability to test many loci with a single primer, and little DNA input. If they only identify the most evident aspects of interest, it may be challenging to reproduce RAPD results from one lab to another. To determine amplification differences, polymerase chain reaction and restriction fragment length polymorphism approaches are employed (AFLP). To synthesize AFLP, PCR-amplified DNA fragments are digested with restriction enzymes at or near a known recognition site in the nucleotide sequence. Individuals from a fast-segregating population with discovered AFLPs may frequently be studied to find polymorphisms and establish linkages. The existence of short oligonucleotide repeat sequences in plant genomes is required for the SSR, STS, SCAR, EST-SSR, and SNP molecular marker families.

6.5: THE REPEAT OF SIMPLE SEQUENCE ALSO KNOWN AS MICROSATELLITE

Microsatellites are small tandem repetitions in the DNA molecule that range in length from 1 to 10 nucleotides. They are also known as simple sequence repeats, small tandem repeats, and simple sequence length polymorphisms. In the scientific literature, the number of PB repeats in microsatellites is frequently given as 2-8, 1-6, or 1-5. SSRs are abundant in eukaryotic genomes and are highly variable and extensively dispersed. The number of repeating units in crop species varies greatly. The repetitive sequence is usually just two, three, or four nucleotides long. Most microsatellites' overall number of repetitions ranges between 10 and 100 dinucleotide repeats. When there are ten or more tandem repeats, there is generally a lot of intra- and interspecific variation in these markers. The PCR is used to amplify SSRs by annealing forward and reverse primers to the 5' and 3' ends of the template DNA, respectively. These polymorphisms can be identified by creating PCR primers for the DNA surrounding the microsatellite region. Even at higher taxonomic levels, flanking zones within a species are frequently persistent (Charlesworth, D. 2010).

PCR fragments are regularly separated on polyacrylamide gels using AgNO3 labeling, autoradiography, or fluorescence detection. Agarose gels containing ethidium bromide can also be used when sample allele sizes differ by more than 10 kb. Creating microsatellite primers from scratch for a new species is a significant technological challenge. Several scientists have explored the characteristics of the developed approaches. Multiallelic and codominant loci are prevalent among those discovered. In a thorough evaluation, bands may predominate, emerge, or fade. Because the flanking DNA is more likely to be retained, microsatellite-derived primers can be employed with many species and types. These vital markers are frequently used for individual genotyping, germplasm evaluation, genetic diversity studies, genome mapping, and phylogenetic and evolutionary research because they are easily accessible, amenable to automation, highly polymorphic, and have high analytical resolution and reproducibility. On the other hand, microsatellites need extensive knowledge of DNA sequences to be manufactured and generate inaccurate estimations of genetic structure; hence, they have been created primarily for domesticated rather than wild species.

SSRs based on expressed sequence tags (ESTs) have been created to broaden the pool of possible SSRs. Because ESTs and other DNA sequence data are widely available, data-mining-based synthesis of EST-based SSR markers is quicker, more effective, and less expensive than synthesizing genomic SSRs. When a large number of clones are utilized for sequencing to detect DNA sections containing SSRs, the costly and time-consuming techniques required to generate genomic libraries can be avoided. Even though EST-SSRs are likely to have lower polymorphism than SSR markers derived from genomic libraries, they are only relevant to species for whom a database exists.

SNPs, or single nucleotide polymorphisms, are changes in a population's genetic sequence caused by a single nucleotide alteration. Sequence polymorphisms are changes to a single base. SNPs (single-nucleotide polymorphisms) are common and can be found near or inside the gene of interest. SNPs are present at various locations in several animal and plant genomes. SNPs can be found using DHPLC or microarray techniques. They have the potential to be employed for a wide range of purposes, including rapid crop identification and the construction of ultrahigh-density genetic mapping. They can be used for research into plant agronomic or adaptation features utilizing genetic mapping and association genetics.

Chapter

7 ECONOMIC BENEFITS OF PLANTING TREES

Contents

7.1: INTRODUCTION

Having trees provides numerous economic benefits. Trees can boost property values, retail sales, and road maintenance expenses. Despite significant potential benefits, forest sustainability programs are the most undervalued environmental boosters of the present day. While forests are being destroyed across the globe, their impact has been diminished in favor of costly experimental air purification techniques. Although the majority of their effects are cost savings for air purification and anti-pollution initiatives, they also support millions of jobs globally.

7.2 AIR PURIFICATION

In addition to being inexpensive, trees effectively purify the air. One of the numerous advantages of planting trees is that they remove carbon dioxide from the air and release oxygen. In addition, they serve as particulate filters. As particulate-laden air flows through the trees, dust particles adhere to the leaves and are washed away by rainfall. According to estimates, trees alone in the United States remove 17.4 million tons of air pollution annually. The value of the benefits to human health was $6.8 billion.

Less expensively, a cleaner environment reduces the risk of airborne diseases. Nearly one million people can find relief from asthma and severe respiratory symptoms thanks to trees. Planting trees can save cities millions of dollars on healthcare costs while improving the aesthetics of urban areas. Additionally, beautiful landscapes improve the civic spirit and mental health. (Richards, D. 2018)

Case study: estimating economic and environmental benefits of urban trees in desert regions

Researchers, urban planners, and decision-makers have placed a great deal of emphasis on urban air pollution as a result of the sharp increase in industrial, transportation, and construction activities that harm air quality. In urban areas, this decline in air quality has been shown to negatively affect the environment and human health. One of the proposed mitigation strategies to improve urban air quality is the use of vegetation. The concept of planting trees to reduce air pollution has recently been investigated. It has been demonstrated that urban vegetation improves air quality, reduces the severity of flooding, and increases property values. Trees can improve the quality of the air we breathe by removing gases and particles from the atmosphere and converting carbon dioxide into oxygen through photosynthesis. Although certain tree characteristics can impede airflow and increase air pollution, trees and bushes located near sources of local air pollution can increase air dispersion, thereby improving local air quality. Moreover, trees provide numerous ecosystem services in urban environments, including the removal of other gaseous pollutants, thermal comfort, and temperature regulation through shading. (Whitlow, T. H. 2019). In addition, by removing, emitting, or resuspending particles from leaf or branch surfaces, trees can directly influence the amount of particulate matter (PM) in the atmosphere. Due to the negative effects of both man-made and natural particulate matter on human health, visibility, and solar energy production applications, mitigating particulate matter in urban desert areas with high levels of dust can be particularly important. Greening cities can improve social, economic, and environmental conditions while also adding color to an otherwise monotonous urban landscape. Thus, there is a connection between green infrastructure and human health. Kumar and colleagues illuminated the relationship between green infrastructure, air quality, and human health in a recent review article. To aid policymakers and urban planners in making decisions, the authors provided a comprehensive understanding of the relationships between the three factors. The authors concluded that urban vegetation can provide numerous health benefits, but there is little empirical evidence linking these benefits to the ability of urban vegetation to reduce air pollution. Significant effort is required to establish the policies, design, and engineering standards that will govern its deployment. To maximize the benefits of greening and quantify the socioeconomic and health advantages of green infrastructure, they also emphasized the importance of developing guidelines for urban greening design.

Understanding and quantifying the structure of urban street tree resources is crucial for evaluating their environmental benefits and for basic street tree management. The most effective way to support street tree management is to conduct a comprehensive census of the street tree population, including location, tree type, and other details. In situations where street tree inventories are unavailable or cannot be conducted, a method for street tree sampling was proposed to provide a relatively inexpensive and straightforward technique for determining the overall street tree composition. The method yielded tree counts with a standard error of approximately 10% of the total population estimate. On the other hand, the author has written about the characteristics of roadside vegetation that can improve local air quality. Height, thickness, coverage, porosity, and species characteristics that promote improved air quality were included among the design requirements. To minimize the effects of air pollution near transportation facilities, urban planners and developers must understand how to preserve existing roadside vegetation or create vegetative barriers, making these design specifications vital. The distance between trees, for instance, affects airflow over a larger area. Studies have demonstrated that vegetation with a minimum thickness of 5 m and, in the majority of cases, a thickness of 10 m or more results in lower concentrations of pollutants near roads. (Richards, D. 2018)

Studies have demonstrated that trees have economic and environmental benefits. In 1994, the removal of 1,821 metric tons of air pollutants by trees resulted in a $9.5 million economic benefit for New York City, according to Nowak (2010). The study, which included other American cities, considered air pollution levels, the duration of the leaf-out season, precipitation, and other meteorological factors. From city to city, the rate at which trees removed pollution varied. This study indicates that large, healthy trees with trunk diameters greater than 30 inches remove approximately 70 times more air pollution annually than small, healthy trees with trunk diameters less than 10 inches.

The Forest Service of the United States Department of Agriculture (USDA) developed the i-Tree software package to estimate the air quality and environmental benefits of urban green infrastructure. Numerous studies have documented the use of i-Tree in the United States to calculate tree benefits. However, only a handful of international studies have utilized this model. Selmi and colleagues published a study in 2016 using the i-Tree Eco model to estimate the amount of air pollution that urban trees in Strasbourg, France, can absorb. According to the model, city-managed trees could remove 88 tons of pollutants annually. 14 tons of nitrogen dioxide (NO_2), 12 tons of particulate matter smaller than 10 microns (PM10), 5 tons of PM2.5, 56 tons of ozone (O_3), 1 ton of carbon monoxide (CO), and 1 ton of sulfur dioxide were removed (SO_2).

In addition, the researchers compared the simulated removal rates to the actual emissions in the area. The public trees of Strasbourg removed approximately 7% of PM10 from the air, but their contribution to the removal of other pollutants was relatively insignificant, according to the findings. It is essential to keep in mind that the effect of each species on ozone levels varies. In a fascinating study that compared numerous tree species, it was discovered that distinct tree types have distinct effects on ozone production, concentration, and secondary organic aerosol mass. The results indicated that isoprene emissions from trees such as black locust, European oak, and poplar increase ozone levels. Monoterpene-emitting trees, including beech, magnolia, and wayfaring trees, produce less overall. To ensure that citizens benefit from climate change adaptation strategies and are not negatively impacted by them, future urban planning, including urban greening, must take into account both the positive and negative aspects of tree species. (Whitlow, T. H. 2019).

Few studies have examined the effectiveness of various urban tree species in reducing air pollution in desert cities with high anthropogenic and natural PM concentrations. Different types of plants have varying degrees of ability to improve air quality, as a result of their unique characteristics. For instance, it has been demonstrated that leaf surface structure has a significant effect on PM2.5 capture.

Little research has been conducted to determine which tree species are best suited for use in arid regions of the world, such as the Middle East, to reduce air pollution. To comprehend the air quality and monetary benefits of planting, it is necessary to investigate the various plant species and characteristics native to this region.

Qatar, one of the nations of the Arabian Peninsula, occasionally experiences extreme air pollution in its major cities. In addition to its arid climate, Qatar is distinguished by its poor soil quality and 88% stony terrain. Qatar, on the other hand, possesses a variety of plants that are regarded as both abundant and diverse. Consequently, tree planting in Qatari cities can have a positive economic and environmental impact by enhancing air quality and other ecosystem services. (Whitlow, T. H. 2019)

As is typical in arid regions, the majority of plant species are annuals, ephemerals, and short-lived perennials due to the harsh climate conditions, including high temperatures and infrequent rainfall. Compared to perennials, which include trees, lianes, shrubs, under shrubs, and perennial herbs and account for approximately 40% of all taxa recorded, these plants comprise more than 60% of the flora. There are seven tree species in the State of Qatar which includes:

- *Avicennia marina*,
- Acacia ehrenbergiana
- Acacia tortilis
- *Ziziphus nummularia*
- *Ziziphus spina-christi*; the lianes

- *Cocculus pendulus*,
- *Ephedra foliata* (a gymnosperm).

In addition to the seven naturalized species listed above, the Arabian Peninsula is home to *Phoenix dactylifera*.

7.3 PLANTING TREES GENERATES EMPLOYMENT

Businesses are boosted by trees. Cities have distinct tree maintenance requirements compared to national forests. People are required to water and prune trees in urban areas. Furthermore, planning and optimizing tree placement require the assistance of professionals. The number and type of trees required vary based on the city and city area. Urban planners, ecologists, and arborists can now find employment. These jobs, which are both important and sustainable, are essential to the success of an urban forest's effect on pollution reduction and health promotion.

It is possible to strike a balance between the preservation of forests and the industries they support through careful management. The global lumber industry employs 13.2 million individuals. Nonetheless, a substantial proportion of these jobs primarily involve deforestation. Cities can increase employment opportunities and economic growth by incorporating trees into their urban landscapes. (Richards, D. 2018).

As more cities and nations develop, the demand for lumber will also increase. Currently, however, the industry is destroying forests, which is detrimental to the environment. Sadly, our rainforests are diminishing. It is possible to realize the full economic benefits of tree planting by enhancing forest management techniques, investing in fire and pest control, and developing stringent monitoring systems. It is possible to create an industry that provides as much as it consumes, thereby preventing the extinction of species, destruction of habitat, and climate change.

Case study: Jobs creation in the USA by trees

With so many people still struggling to regain their footing, the importance and value of trees in job creation are greater than ever. Considering that green industries support two million jobs in the United States, including 240 thousand in California alone, one could say that money does indeed grow on trees.

The "green industry," also known as the horticultural industry, consists of arborists, nursery and sod growers, landscape architects, designers, and builders, contractors and maintenance companies, retail garden centers, home centers, and mass merchandisers with lawn and garden departments, as well as marketing intermediaries such as brokers and horticultural distribution companies. These jobs require a wide variety of skills (re-wholesalers). According to the report Economic Impacts of the Green Industry in the United States, this industry is one of the most rapidly expanding in the economy. It is so robust that even during recessions it frequently expands.

Young adults in the United States were among those hardest hit by the recession. Teenagers can use trees and other vegetation as a significant springboard on their way to becoming valuable contributors to our economy. The San Diego County Urban Corps has served in this capacity for the past two decades. (Whitlow, T. H. 2019).

The Corps offers environmental job training and educational opportunities to young adults in fields such as urban forestry and conservation. Corps members work in the community on environmental projects such as tree planting four days per week while attending high school classes one day per week. Urban Corps has helped over 6,500 young adults throughout its 20-year history.

Indirectly, trees support the countless additional jobs required to produce the numerous products that are in some way related to trees. Have you ever wondered where your gum and pencils originate? The trees motivate them to run. Remember the 18-billion-dollar fruit and nut industry, which includes apples, pears, almonds, and hazelnuts.

According to data from 2004, green industry sales totaled $147 billion, with $28 billion coming from our western region. Trees are undeniably a wise investment, given that they generate tens of thousands of jobs and billions of dollars in revenue. The bottom line is that the expansion of our economy parallels the growth of trees!

How reforestation can help combat the unemployment crisis in the United States

The motto of the California Conservation Corps appears to be more of a warning than a recruitment tool, but for thousands of individuals, it holds the promise of new skills, a career start, and even a bit of adventure.

In addition, it accurately depicts the work of the approximately 3,400 young people trained annually by the state agency for employment in forestry and other conservation-related fields.

It takes a great deal of effort to plant tree seedlings on steep summer slopes and to clear dense vegetation from forests so that existing trees can grow.

Growing and maintaining trees in urban areas can be challenging. Digging large holes for resilient trees that can withstand the relatively harsh conditions of urban environments and climbing trees to prune branches so that they do not fall on cars and houses can be exhausting.

As the nation's interest in trees as a solution to climate change and social injustice increases, we urgently need more people to perform this type of work. (Richards, D. 2018)

Despite the difficulty of the task, the timing of the opportunity could not have been better. One out of every ten Americans was unemployed in August. People from underprivileged urban and rural communities, which have the greatest potential for forestry employment, have been among the hardest hit by the recession.

The World Economic Forum's global initiative (1t.org) to conserve, restore, and cultivate 1 trillion trees by 2030 presents a significant opportunity for the creation of forestry jobs. As part of the movement, nearly 850 million trees will be planted this summer by 28 businesses, government agencies, and nonprofit organizations in the United States. The World Economic Forum and American Forests are two of the most influential organizations in the United States. They aid 1t.org chapters in fulfilling their pledge obligations.

The devastation caused by climate change-related wildfires, droughts, pests, and diseases in the national forests of the United States is a related pressing need. This enables the employment of a large number of individuals. There is already a backlog of 7,7 million acres of land that needs revitalization. Frequently, addressing this need will necessitate a combination of actions, such as planting trees and thinning dense forests, to increase their resilience and diversity and promote the growth of new trees. The backlog is especially severe due to drastic changes like the work being performed and reductions in the federal forestry workforce as a whole. Especially in the western United States, more federal foresters are fighting wildfires than caring for existing trees and thinning the forest to remove fire fuels.

The need for tree equity in cities also presents an opportunity to employ unemployed urban residents to plant and maintain trees in areas where they are most needed. Each neighborhood in a city must have a sufficient number of trees so that everyone can reap their benefits. In the majority of cities, low-income and minority neighborhoods have few trees. More than 20 million acres of public land in the United States have been identified by the Nature Conservancy as having reforestation potential. To restore the forest on this scale, 6 billion trees would need to be planted, requiring 582,000 people to work as tree planters, forestry professionals, and conservationists, among other positions. The total expense was roughly $14.66 billion.

Jad Daley, president and chief executive officer of American Forests, states, "We have a tremendous need to restore our forests and plant billions more trees in America."

Given the economic downturn, this is an excellent opportunity to employ large numbers of urban and rural individuals. We need to mobilize the populace. To achieve this goal, resources and creative ideas to make forestry jobs appealing and accessible to all will be necessary.

Planned reforestation of millions of acres in the United States will prioritize entry-level tree planting positions. However, filling these positions can be challenging. In addition to being physically demanding (workers in large forests frequently plant hundreds of trees per day), it is also poorly compensated. The average hourly wage for forest or conservation work in the United States was $15.76 per hour in 2016, with tree planters earning the lowest hourly wages in the forestry industry.

In addition, the jobs are seasonal and last only a few months. Seasonal workers comprise the majority of those who plant trees in the United States. Other seasonal forestry occupations include fighting wildfires, planting trees in nurseries, and managing forests with prescribed burns. Consequently, attracting and retaining individuals in the field may be challenging.

If we want to attract more workers, have a lasting impact on people's lives, and stimulate the economy, we must create a large number of stable, long-term jobs with advancement opportunities. (Whitlow, T. H. 2019)

According to Zander Evans, executive director of the Forest Stewards Guild, a professional association for foresters and other land stewards based in Santa Fe that also offers training programs, combining seasonal jobs to create positions that are more stable and last for longer periods is one way to create a stable workforce. Clearing fire-prone underbrush and suppressing wildfires could be combined with nursery and planting work. This would necessitate the teaching of multiple skill sets to individuals.

Evans asserts, "We are all aware that planting trees for eight to ten hours per day is laborious." However, I believe it would be beneficial to pay those who cultivate and properly plant trees a living wage. However, the fragmentation of the forestry industry makes it difficult to design these positions. Numerous public and private organizations and individuals, including tree planters and other seasonal employees, work in forestry. To mobilize a vast army of more robust and stable positions with advancement opportunities, government and private employers would need to work together.

Brian Kittler, senior director of forest restoration at American Forests, explains, “Creating this kind of capacity requires tying together through relationships the workforce we’ll require.”

As a result of recognizing this need, leaders from various forest sectors have developed strategies and policies over the past few years to better leverage the knowledgeable forestry staff in state and federal agencies with additional capacity from non-governmental organizations and forestry contractors.

This is an important aspect of the relatively new concept of “shared stewardship,” which aims to make it easier for governments, nonprofits, tribes, and private businesses to collaborate and pool resources in initiatives that cross forest ownership boundaries. Similarly, the Forest Stewards Guild’s trained and organized workforce provides essential labor for tree planting, prescribed fire, restoration thinning, and wildland firefighting across the shared landscapes of New Mexico. American Forests collaborates with several states to advance shared stewardship strategies, including workforce assessment, as restoration work expands.

The method could have a significant impact on rural communities as a result of extensive planting and forest restoration. Before the COVID-19 pandemic, which devastated their economies, many rural communities had long-standing economic problems. These regions would benefit economically from stable, year-round employment in forest-related industries, such as the manufacture of wood products or the provision of services to the massive outdoor recreation industry. (Richards, D. 2018)

7.4 TREES INCREASE PROPERTY VALUE

The majority of mature, healthy trees will increase the value of a property, which is excellent news for gardeners. According to Kathleen Wolf, a social scientist, in an article for Arborist News, homes with trees are generally preferred over comparable homes without trees, resulting in a 7% price increase. The situation becomes more complex when attempting to identify the specific species that will contribute the most value. According to Wolf, it is effective to assign values to trees sold as home components using "hedonic pricing," or pricing based on indirect or subjective preferences that cause people to pay more.

She then reveals an astounding fact: mature trees in affluent neighborhoods typically increase a home's value by 10 to 15 percent, which is significantly higher than the norm in less affluent neighborhoods. Therefore, if you live in a region where homes are generally more expensive, the addition of a mature tree will increase the value of your property even further. In less expensive areas, the price increase may not exceed 2%.

Trees increase the value of your property, add aesthetic appeal, and can reduce your cooling and heating costs. According to the American Power Association, effective landscaping can reduce a home's annual cooling costs by as much as 50 percent. It is suggested that you plant trees on the north and northwest sides of your property to create a natural barrier against winter winds and reduce heating costs. (Whitlow, T. H. 2019)

How do you choose a tree to increase your property's value?

Some homeowners have difficulty choosing which trees to plant, especially if they are unfamiliar with them. The success of the tree and the prevention of further damage depends heavily on choosing the proper tree for the appropriate location. An arborist recommends the following steps for selecting the ideal tree for your home.

- Choose a tree species that thrive in your region's cold climate.
- Select a tree with the qualities that are most important to you, such as a large canopy for shade, rapid growth, color, screening, etc.
- The planting site's soil will determine which tree species will thrive there. Parker states, "You must have a properly prepared site." In this instance, the condition of the soil is crucial.

While the majority of trees favor full sun, a few require partial shade. The quantity and quality of available light at the planting site will determine the optimal species.

- Like other plants, trees require sufficient spacing. Significantly contributing to the decline of the landscape is overplanting. Parker recommends "planning for the tree's growth over the next ten, twenty, or thirty years."

The diversity of plant species is essential in a landscape. Avoid concentrating on a single tree species. When unsure of which species to select, choose a native species. "Native plants generally grow well and require little upkeep when grown on proper soils in the proper environmental conditions," says Parker. By choosing the appropriate native plants, it may be possible to reduce pesticide and water usage.

Some of the best tree varieties to increase property value include:

- An oak tree

Long-lived oaks are popular trees for residential landscapes due to their beautiful large canopies, rounded spread, stunning fall foliage colors, and capacity to support more species than any other tree in North America.

If you're looking for a tree to provide summertime shade and keep your yard cool, look no further. The Magnolia cucumber or *Magnolia acuminata* tree grows rapidly, produces cucumber-like fruit, and bears greenish-yellow, faintly fragrant flowers.

- Sugar Maple (*Acer saccharum*)

This tree's thick crown provides excellent shade during the spring and summer. Expect yellow, burnt orange, and red foliage in the fall. In the spring, it bears tiny, greenish-yellow flowers.

- Tupelo (*Nyssa sylvatica*)

The autumn ripening of this tree's small, bluish-black fruit attracts a variety of birds and other wildlife to your yard. It puts on a show in the fall with leaves in various shades of yellow, bright red, purple, and scarlet. The trunk develops bark that resembles alligator skin as it ages. Other trees, such as the Southern Magnolia and the Red Maple, will undoubtedly increase the value of your property, in addition to those already mentioned.

7.5 PROTECT THE INFRASTRUCTURE

By providing shade, trees prolong the life of municipally owned assets and infrastructures, such as roads and footpaths. The shade provided by trees extends the lifespan of the pavement. According to the results of a California study that examined the daily heating and cooling of asphalt, street trees can increase asphalt's lifespan by 40 to 60 percent. Due to the high cost of asphalt, this will help make the road system more accessible. (Richards, D. 2018).

7.6 TREES INCREASE BUSINESS VALUE

Various surveys have found that businesses on tree-lined roads experience an average 20% increase in revenue. This characteristic makes main street stores more successful than mall stores.

To increase sales, a growing number of businesses are placing a greater emphasis on establishing themselves in locations with distinctive scenery.

Trees benefit businesses! A recent study found that shoppers prefer shopping districts with a healthy urban forest. Many revitalizing business districts across the nation are working hard to create vibrant, vital consumer environments. Why should trees be incorporated into an action plan? Healthy, well-maintained trees convey positive messages about the allure of a district, the quality of its goods, and the level of customer service one can anticipate. They are an essential component of any strategy for attracting customers and visitors. To revitalize neighborhoods, it is necessary to address pressing issues such as security, sanitation, parking, and marketing. Any improvement program must give trees top priority. Moreover, the favorable environment created by trees may even contribute to the resolution of some of the remaining issues. The majority of retail environments in the United States have a tree canopy cover of 5% or less; American Forests, a national tree nonprofit, recommends a 15% canopy cover goal for business districts. According to studies, investing in trees will improve the bottom line of your business!

Case study: Research Project

The national study conducted by the University of Washington utilized survey questionnaires to determine how individuals felt about trees' contribution to commercial resurgence. Districts in the Pacific Northwest, Austin, Los Angeles, Chicago, Pittsburgh, and Washington, D.C. received questionnaires. In a survey, the responses of business owners and managers were compared to those of locals, potential customers, and shoppers.

Our indoor and outdoor environments influence how we spend our days. (Whitlow, T. H. 2019). The physical characteristics of a location affect how we navigate and move through it. A location's environment also sends subliminal cues that influence our attitudes and behavior there. This study examined how the character of a place affects how people respond to a business district. A series of questions were posed to participants regarding their preferences and behavior in three fictitious business districts. The key findings of the study are summarized below.

Perception of place: The ratings of the three business districts by survey respondents revealed four distinct perceptions: Amenities and convenience Interaction with Professionals in Business The Goods' Specification Maintenance and repair In every category, consumers rated districts with street trees and other landscape enhancements significantly higher! A tree-lined sidewalk, for instance, received approximately 80% higher Amenity and Comfort ratings than an unshaded street. Moreover, districts with trees performed 30% better in terms of Product Quality than those with bare sidewalks. Customer service issues were among the interactions with merchant products; in districts with trees, ratings were 15% higher.

Patronage behavior: The way we feel about a location influences our choices. Participants were asked about their travel patterns between the three shopping districts, as well as their willingness to pay for parking and their travel time and distance. Once again, trees are essential! When all behaviors were considered, districts with trees reported higher measures. For instance, respondents indicated that they would be willing to pay more for parking in a commercial district with attractive landscaping. This implies that increased parking revenue would more than compensate for the loss of parking spaces, a common complaint business owners have about trees.

Pricing Patterns: Do trees affect consumers' willingness to pay for a product? Utilizing contingent valuation techniques, the relationship between amenity values and customer price valuations was determined. In the business districts, fifteen items were placed in a "basket of goods," and survey participants were asked to provide a price for each item. The merchandise was categorized as convenience, shopping, and specialty. Respondents to the survey consistently assigned significantly higher prices to goods in landscaping districts! The average price of goods in the landscaped area was 11% higher than in the unlandscaped area. This was true for both expensive, shopped-for items and inexpensive, quick-purchase convenience items (such as a sandwich or a bouquet) (e.g. sports shoes, new glasses). Considering the razor-thin profit margins of the majority of retail establishments, trees appear to offer a substantial "amenity margin."

7.7 REDUCING ENERGY COSTS

The shade that trees provide helps to cool the environment. Temperature plays a significant role in determining energy costs. In the shaded neighborhood, households can reduce their energy costs by 15 to 35 percent. A 25-foot tree reduces annual heating and cooling costs by 8 to 12 percent, saving the average American family $10 annually. In addition, urban structures and pavement generate a "heat island" effect. A mature tree canopy reduces air temperatures by 5 to 10 degrees Fahrenheit, influencing the interior temperatures of nearby structures.

Insulation and weatherproof windows and doors are two energy-saving home improvements that can be equally as effective as strategically placed trees. The presence of trees can reduce your heating and cooling expenses.

By cooling the air in the summer and blocking the wind in the winter, trees conserve energy. Consequently, fewer fossil fuels are used to generate electricity for heating and cooling using fewer fossil fuels.

A minimum of three large shade trees strategically placed around a home can reduce cooling costs by up to 30 percent. The greatest advantages of shade trees are realized when:

- Plant deciduous trees, whose leaves fall off in winter. In the summer, these trees provide shade and block the sun's rays; in the fall, they shed their leaves to let in light.

Plant these trees on the west and south sides of the structures.

To reduce the heat load on the landscape, shade all hard surfaces such as driveways, patios, and walkways.

Utilize evergreen arrangements that retain their leaves and/or needles throughout the year. As windbreaks, they will reduce heating energy requirements by 10% to 50%. The greatest benefits of evergreens occur when:

- Position them on the north side of your house to deflect and moderate winter winds.
- Do not plant them on the south or west sides of your home, as this will prevent the winter sun from warming your home. In the summer, these trees also provide some shade.

By lowering air temperature, increasing humidity, influencing wind speeds, and reducing glare, trees contribute to local climate change. When trees are thoughtfully planted along streets and near buildings, urban areas experience cooler summer air temperatures. These trees provide both shade and evaporation of water through transpiration. A single large tree can evaporate as much water as ten rooms' worth of continuously operating air conditioners. Communities benefit when homeowners include trees in their landscaping. The proper placement of trees in landscaping helps homeowners save money on daytime cooling costs. For instance, strategically placing three or more large trees on the sunny sides of a home will provide enough shade to reduce the cost of air conditioning by up to 30 percent. In the winter, trees aid in energy conservation by blocking cold winds and reducing the strain on heating systems. When applied to a large number of homes, communities, and cities, these energy savings can reduce the demand for electricity from utility plants, thereby reducing the number of air pollutants these plants produce.

7.8 THE PRESENCE OF SHADE SAVES MONEY

Did you know that shading just 17% of your home during the day can save you $10 per month on your electricity bill? In addition, reducing the same shade to 50% will save you an additional $20 per month on your energy bill. It requires time to plant a tree and provide shade for those who lack it. According to one study, energy savings can begin five years after planting and reach 12% after fifteen years. Continue reading to discover the key to success, whether you already have tree shade or wish to create it: Place the suitable tree in the suitable location.

7.9 THE COOLING EFFECT OF TREES

Trees have a naturally occurring cooling effect on the environment. Through transpiration, which is similar to evaporation, trees lose water vapor. This allows minerals and water to circulate throughout the tree as well as cools the tree and its surroundings. It is comparable to how sweat cools our skin.

According to one study, the net cooling effect of a young, healthy tree is equivalent to that of ten room-size air conditioners operating 20 hours per day. In urban environments, concrete, asphalt, and other impervious surfaces absorb heat, causing heat to accumulate and dissipate well after sunset. Consequently, cities are typically several degrees warmer than their surrounding rural areas. The term for this is the Heat Island Effect. There is however hope for the urban environment if trees are introduced and utilized. In addition to lowering urban temperatures, urban trees can reduce winter heating and cooling costs.

Planting trees for summer shade

The sun's rays heat home surfaces such as walls, roofs, and windows, causing air conditioners to work harder. The result is an increase in energy consumption and utility costs. The shade provided by nearby trees can further reduce energy consumption, particularly in older homes with insufficient or obsolete insulation.

For maximum energy efficiency, choose deciduous trees, or trees that lose their leaves in the winter. During the cold winter months, when the sun's warmth is preferred, these trees will provide little shade. The key to maximizing energy conservation is positioning trees so that they shield the house from the morning and afternoon sun.

Plant trees to provide morning and afternoon shade for east and west-facing walls and windows. In both the east and the west, a variety of small and large trees should be planted. When the sun is low on the horizon in the early morning and late afternoon, small trees will provide shade. In contrast, larger trees that mature to a height of at least 25 feet will provide shade in the morning and late afternoon when the sun is higher in the sky.

Additionally, trees should be planted on the south and southwest sides of the home to provide summertime roof shading. The crowns of these trees should be tall and broad enough to reach the roof. Smaller trees or trees with an excessive number of lower branches, even if deciduous, should not be planted here because they will cast too much shade in the winter.

Planting trees to provide shade for the air conditioner is another method of conserving energy. To facilitate airflow, trees and their branches must be kept at least a few feet away from the units. If this flow is restricted, the device will operate less efficiently.

Plant larger trees at least fifty percent of their mature width away from your home's foundation to prevent conflicts and damage. Plant a tree at least 20 feet from your home if its expected mature width is 40 feet. Smaller trees and shrubs that are less likely to cause conflict can be planted closer to the house to provide shade until the larger trees mature and begin providing shade.

Planting for winter energy conservation

The most significant contribution trees can make to winter energy conservation is blocking cold winds. In the winter, winds can force cold air through cracks, holes, and frequently opened doors and walls. According to estimates, home heating accounts for 53% of total energy consumption in the United States.

A properly constructed windbreak can reduce heating costs by 10 to 25 percent. Windbreaks function by reducing wind speed and redirecting it over buildings. It is best to plant windbreaks perpendicular to the direction of winter winds on the west, northwest, and north sides of the house. Local weather patterns and conditions must be taken into account when determining the optimal planting locations.

Between 2 and 5 times the height of the windbreak downwind, the wind is reduced. This means that if you want your windbreak trees to grow to a height of 30 feet, you should plant them 60 to 150 feet away from the structure they are meant to protect. This may not be possible in smaller yards, but the breaks will continue to function, albeit to a lesser extent.

For effective windbreaks, select evergreen trees with low-lying crowns (meaning they keep their leaves all winter). These trees should be planted close together so that their crowns will intersect. This distance may be as little as 6 feet for trees with smaller or narrower crowns and as much as 15 feet for trees with larger or wider crowns. Planting patterns may consist of clusters, curved lines, or both. Groupings are recommended if you live in an area prone to strong winds, such as hurricanes.

7.10 TREES AND FOOD PROVISION

In numerous ways, forests and other tree-based ecosystems, such as agroforestry, contribute to food and nutritional security. Trees provide an assortment of nutritious foods, including fruits, leafy vegetables, nuts, seeds, and edible oils, which can help to diversify diets and fill nutritional gaps caused by seasonal food shortages. The diversity of edible plants, fungi, bush meat, fish, and insects is also greater in forests. In addition to providing "green fertilizer," fodder for meat and dairy animals, and wood fuel, which is essential in many communities for cooking food, tree-based systems support crop production. Forests and other tree-based ecosystems indirectly help communities acquire food by generating income, and they also provide environmental services that support crop production. It is difficult to calculate the relative advantages and disadvantages of food production systems based on trees. It is frequently difficult to comprehend how tree-based systems function due to these obstacles. A deeper understanding focuses on critical indicators, such as measures of dietary diversity, as well as systematic methods for characterizing effects across diverse landscapes. (Whitlow, T. H. 2019)

Trees bearing fruit

This is the most well-known type of tree food, and it includes mangoes, avocados, apples, oranges, and bananas, among others. These fruits are consumed worldwide and are rich in vitamins A and C, which are frequently lacking in the diets of many people. Fruits with orange flesh are rich in vitamin A and simple to cultivate in tropical climates. Wild fruit varieties are also valuable.

"Fruit trees can provide year-round products for consumption and sale if farms cultivate groups of species with varying harvest times," she continued.

Numerous trees produce edible nuts and seeds that provide essential dietary fats and oils.

Dietary products derived from tree leaves and stems

Numerous individuals consume both cultivated and wild tree leaves. They can be used fresh, dried, powdered, or fermented for preservation.

Trees tubers and roots.

Due to their ability to withstand low rainfall, roots and tubers provide carbohydrates and are a significant source of water during drought or famine.

Trees for animal consumption

For the production of food for millions of people, a sufficient supply of livestock forage is essential. Farmers who raise animals on arid and semiarid lands frequently feed their animals tree twigs, leaves, small branches, seedpods, and fruit.

In some instances, trees are grown solely for their fodder, and the foliage is hand-harvested to feed animals in stalls. Increasingly, livestock is permitted to browse naturally growing trees and shrubs. Whether harvested or browsed, forested forage supports livestock production and ensures the year-round availability of dairy and meat products.

The nectar of a flowering tree

Bees are nourished by the nectar of flowers on trees such as eucalyptus, orange, chestnut, and acacia, which contributes to the maintenance of the natural cycle and the production of honey for human consumption.

(Whitlow, T. H. 2019)

Chapter

8 COMMODITY GROUPS OF FOOD AND OTHER ECONOMIC CROPS

Contents

8.1 INTRODUCTION

Crops and animals that are raised on farms or plantations are considered agricultural commodities. The majority of agricultural products, including grains, livestock, and dairy, supply food to people and other animals all over the world.

Almost all living things on Earth are reliant on the agricultural sector in some way. Farmers produce grains, fruits, vegetables, and livestock. We use lumber to build our homes. We dress in cotton and wool. And we have rubber tires on our cars. (Magdoff, F. 2012)

8.2 ROOTS

Root and tuber crops are any plants that produce starchy stems, rhizomes, corms, roots, or tubers. They are generally utilized in the manufacture of starch, alcohol, and fermented beverages like beer in addition to human food (either raw or processed) and animal feed. For many of the poorest farmers and undernourished households in developing countries, root and tuber crops represent a vital source of food and nutrition. Root and tuber crops create a lot more energy each day than grains do. They now play a significant role in maintaining the food and energy security of the entire planet. The principal root and tuber crops are potato, sweet potato, cassava, and yam, which cover roughly 53.93 million hectares globally and yield 736.747 million tons yearly. Potato, sweet potato, cassava, and yam are some of the most significant food crops in the world in terms of yearly output volume. Three of the top ten food crops farmed in developing nations are cassava, sweet potatoes, and potatoes. The origin, distribution, botany, physiology, breeding, cultivation, pests, and illnesses, as well as the uses and economic importance of three crops—potato, sweet potato, and cassava—are addressed in this section. (Borer, H. 2014).

RTCs are the second-most major source of carbohydrates for humans and the most significant in Sub-Saharan Africa because they produce a range of edible belowground organs. The production of adventitious roots (ARs), storage root differentiation, and the growth and breakdown of non-storage roots are the key issues of this paper. Although the storage organs of RTCs and ARs are similar, their root systems are separate, and it can be misleading because a lot of the literature refers to them all as tubers. In cassava and sweet potato, SR development is triggered by root cambium growth and proliferation of starch storage tissue, with highly expressed lignin biosynthesis genes and downregulated lignin biosynthesis genes. In the beginning and development of storage organs in cassava, potato, and sweet potato, KNOX1 activity is related to a variety of genes, including two MADS-box transcription factors. Generalization is problematic owing to the small number of studies, however, some suggest that storage roots/tubers begin to develop quickly at the same time as they reach their maximum length. Variations in soil water and nutrient availability may be reflected in distinct growth patterns. Many ARs seem to be transient occurrences. While cassava had a larger rooting depth (> 1 m) but a lower root length (1-2.84 km/m), potatoes generally had a rooting depth of 1 m and a maximum root length of 7–12 km/m2. The necessity to quantify the lifetime and usefulness of diverse root types, notably SRs, as well as the scarcity of RTC research and the uneven language used to describe roots and storage organs are also noted. (Magdoff, F. 2012)

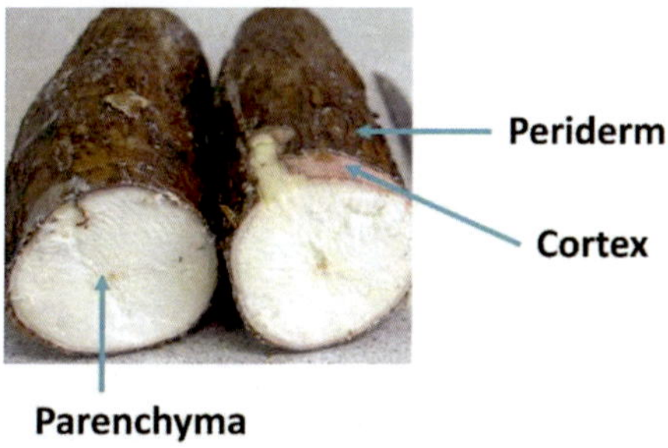

Figure 8.1: Intact root showing different components.

Source: Tumwesigye, K. S., Oliveira, J. C., Namuwaya, S., & Sousa-Gallagher, M. J. (2021). Cassava Biomaterial Innovations for Industry Applications. In (Ed.), Cassava - Biology, Production, and Use. IntechOpen. https://doi.org/10.5772/intechopen.97493

Small-scale farmers in the area plant root and tuber crops such as cassava, sweet potato, and yams, which give reasonable yields despite the area's recurring climatic change and fluctuation as well as its substandard agroecological conditions. Compared to other staples, root and tuber crops give greater energy and nutrition per unit of land area and per unit of time, with certain cultivars supplying large amounts of iron, zinc, calcium, and vitamins A and C to diets. They may be turned into a range of food, feed, and industrial goods that are prepared for consumption or sale in the house and require less intensive management methods. They serve a crucial role in the yearly cycle of food as a safety net, ensuring food availability and security in families, particularly those with a limited labor supply owing to HIV/

AIDS, because of their flexibility, different maturation duration, and in-ground storage potential. They are amenable to growing in a mixed-farming system, contributing to food diversification through time, and supplying food during times when people are most in need, minimizing the impact of grain scarcity. They are also a key source of employment and income for women, particularly living in rural, sometimes underserved areas. They also have substantial economic potential as suppliers of raw materials for the starch, feed, and biofuel businesses and markets.

For individuals in the developing world who are most in danger from hunger, root and tuber crops supply a core component of their nutrition. Food staples are foods that are eaten often in big quantities and are a substantial source of nutrients and energy in the diet. While potatoes fall behind the three primary cereal grains of maize, wheat, and rice, four of the top 10 dietary staples globally are root and tuber crops. In many resource-poor and developing countries, food farming requires crops that can grow in marginal soils and are supplemented with minimum inputs. Except for potatoes, practically all root and tuber crops are cultivated in the dramatically variable climes of the tropics.

Industrial production and processing of root and tuber crops are problematic owing to the characteristic and usually uneven physiological form of the roots and tubers generated. Differences in root and tuber morphology can be ascribed to the diverse conditions in which they are produced. Though their application in large-scale commercial production may be hampered by the same then plasticity that allows these crops to flourish in a range of conditions, this adaptive attribute is vital for feeding the needy in underdeveloped nations.

Crop breeding focuses on choosing hybrids for desirable features. To generate children who outperform either parent, crosses are done between parental lines that have varied and presumably complementary genes. The production of new cultivars is tough since breeding needs a lot of time and effort. Breeders must keep up with quickly changing plant infections capable of damaging the essential harvest, which makes vegetative propagation risky, as in the case of root and tuber crops. Plant diseases may impact a whole crop with minimal genetic variation, thus farmers are defenseless against them.

The potato is the most productive crop in the world in terms of input needs to total caloric output. With an annual global production of more than 335 million metric tons, potato consumption has more than quadrupled since

1993. In terms of worldwide potato output, China is in the lead, followed by Russia and India. Large-scale potato cultivars often have shallow root systems that make harvesting simpler but restrict their access to lower soil profiles that store water and nutrients. The potato tuber is an expanded part of the stolon that grows in response to hormones caused by temperature and day length. Tubers develop swiftly once they start, yet extreme temperatures can hinder formation. (Magdoff, F. 2012)

To increase water drainage and irrigation consistency, potatoes are planted commercially in hills or raised beds. Potatoes may be cultivated on a range of soils with varying amounts of water retention. Ideal soils would have adequate structure and drainage, allow for root aeration, and decrease the impacts of root disease. Although soil pH may range from 4.5 to 8.5, 5.5 to 7.0 is optimum for growth and nutrient absorption.

In North America, the planting season for potatoes starts in March or April with the preparation of the seed. The harvest begins in late July, while the growing season spans from April through October. 15–30 days after planting, the first symptoms of sprouting and emergence will show. Tuber initiation occurs 15–30 days after emergence; stolons begin at the same time. 45 to 90 days after emergence, tubers start to fill. Between 90 and 120 days following emergence, the plant will achieve maturity, grow tubers, and be ready for harvest.

If not appropriately treated, potato fungal infections including early blight (*Alternaria solani*) and late blight (*Phytophthora infestans*) can severely influence crop productivity. To reduce crop loss, further pest management activities, such as aphid and psyllid control, are necessary. Nematodes can have a similarly devastating influence on total productivity if they are not managed by basic crop rotation.

8.3 CEREALS

Any grass (family Poaceae) with edible, starchy seeds is considered cereal. Generally speaking, the nutritional value of grains is about the same as that of other foods high in carbohydrates but low in protein, calcium, and vitamin A. It is common practice to add extra nutrients to bread, especially those made with refined flour, to make up for any deficiencies in the grain. Wheat, rice, rye, oats, barley, corn (maize), and sorghum are the most widely cultivated grains. Cereals are commonly distributed as raw grain (though they may also be found in frozen or canned forms) or as components of a wide range of meal preparations. Cattle and fowl eat them as animal feed, and then humans eat the meat, dairy, and poultry that the animals produced. Industrially, many grains are used to produce sugar, glue, oils, and alcohol. (Choudhary, A. 2020)

Native Americans domesticated corn (*Zea mays*) in the Western Hemisphere before early travelers brought it to Europe. The United States is the largest producer, although it is a staple crop in most temperate regions. Maize is sold as a fresh vegetable, processed into canned or frozen goods, and used by humans. Many other culinary products, such as corn flour, corn oil, and corn syrup, are made from the grain as well. It is widely used in the production of cellulosic ethanol, a biofuel, and as an extremely important animal feed.

The Asian diet centers around rice (*Oryza sativa*), which is the world's second-most-produced grain crop. Rice is often grown in small paddies and harvested by hand, as opposed to wheat, which is typically grown in vast areas and harvested mechanically. The paddies are filled with water, often up to around 15 centimeters (6 inches), and then drained and dried right before harvest. The vast majority of milled rice is used in the immediate community. Rice is used in a variety of other products, including breakfast cereals and alcoholic drinks like Japanese sake.

Wheat (different species of Triticum) is a staple cereal crop and one of the earliest domesticated cereal grains. Meal, breakfast cereal, and flour for baked items are all made from wheat today. It's adaptable to a broad variety of soil types, although it does best in temperate zones.

Rye (*Secale cereale*) is second only to wheat in popularity for use in baking. It's also a common ingredient in a variety of baked goods and distilled beverages, especially whiskey. Rye is a hardy grain that may be grown in less-than-ideal conditions and can weather harsher winters than most other cereals. When it comes to rye production, Poland and Russia are two of the world's top countries.

The Avena sativa plant, from which oats are derived, is cultivated around the globe wherever there is a temperate climate, but especially in the United States, Canada, and northern Europe. Animal feed accounts for the vast majority of the demand for oats, although some of the crops are also used for human use. In milder regions, farmers also sow barley (*Hordeum vulgare*). It may grow in less nutrient-dense soils as well. Malt for beer and whiskey is derived from barley, which is mostly used for feeding livestock. Choudhary, A. (2020). Besides being a key ingredient in beer, barley malt may also be found in a wide variety of vinegar and morning meals. (Magdoff, F. 2012)

Milo is the popular name for the sorghum (*Sorghum vulgare*) plant, which is mostly produced for its usage as animal feed. Numerous countries cultivate teff (*Eragrostis tef*) and millet (numerous species) for local consumption and export, respectively.

Sorghum (Sorghum bicolor) is a cereal grain plant of the grass family (Poaceae) with delectable starchy seeds also known as enormous millet, Indian millet, milo, durra, and shall. Grain sorghums, used for food; grass sorghums, produced for hay and fodder; and broomcorn, used in producing brooms and brushes, are only some of the varieties of this plant that have been domesticated since its presumed African origin. Sorghum is called jowar, cholam, or jonna in India; Guinea corn in West Africa; and kaoliang in China. Sorghum's great demand in arid regions is due to its ability to withstand both heat and drought.

Sorghum is a robust grass that typically reaches heights of 0.6–2.4 meters (2–8 feet) but can grow as tall as 4.6 meters (15 feet) in exceptional circumstances (15 feet). The pith, or internal component, of the stalks of some types, is juicy and tasty, while the stalks themselves are covered with white wax. There is about a 5 cm (2 inch) width and a 76 cm (2.5 foot) length to each leaf. Each cluster of these tiny flowers, called a "panicle," contains between 800 and 3,000 kernels. The seeds are smaller than wheat seeds but vary greatly in color, shape, and size according to the variety.

When compared to corn, the feed quality of sorghum is lower (maize). It contains calcium, iron, vitamin B1, and niacin in trace amounts, but is mostly composed of carbs (92%). It is common practice to mill gluten-free grain into a meal before using it to make porridge, flatbreads, and cakes for human consumption. Processing might lessen the typically robust taste. Oil, starch, dextrose (a sugar), paste, and alcoholic drinks are all products of the grain. Feed and structural components are made from the stalks. Grazing animals, syrup, and ethyl alcohol for biofuel are the most common uses for the sweet sorghums (also known as sorgos) that are mostly cultivated in the United States and southern Africa.

Common buckwheat, or buckwheat (*Fagopyrum esculentum*), is a member of the Polygonaceae family of plants that are grown for its edible seeds. The hulled kernels, or groats, of buckwheat, are prepared as kasha, boiled, and consumed much like rice, making it a substantial pseudo-grain crop in certain regions of eastern Europe. Although buckwheat flour is not suitable for making bread, it is used in the United States and Canada to make a type of griddle cake known as buckwheat cakes. Buckwheat is high in both carbs and protein, with only trace amounts of B vitamins 1 and 2. (Choudhary, A. 2020)

Green buckwheat (*Fagopyrum tataricum*), a related plant, is also widely cultivated in East Asia and put to similar uses. Buckwheat is an annual plant with rapid growth. The hollow crimson stalks support the heart-shaped leaves, which are quite basic. Flowers of a pure white hue are produced in terminal clusters, where they attract pollinators like bees and butterflies. A thick dark brown or gray skin protects the kernels of these triangular-shaped seeds. (Magdoff, F. 2012).

Although buckwheat yields are lower than those of actual grain crops on rich soils, it thrives in the harsh conditions of mountainous, desert areas and cold climates. It may be grown as a late-season crop due to its rapid growth. Honey may be made from buckwheat flowers because they provide both pollen and nectar for bees. The plants can be sown as a green manure crop and then plowed under to improve the soil, which allows for better circumstances for the growth of other crops by burying weeds and attracting beneficial insects. It's commonly used to feed chickens and other livestock, and it's particularly popular in England for usage with pheasants.

Green algae belonging to the family Ulvaceae, commonly known as "sea lettuce" (genus Ulva), can be found growing on the stony beaches of oceans and seas across the world. In addition, certain species thrive in brackish water that is high in organic matter or sewage and can accumulate heavy metals. *Ulva lactuca*, a kind of sea lettuce, is used as a salad addition in northern Europe, Japan, and China due to its high iodine and vitamin A, B, and C content.

The thallus is a sheet of cells up to 30 cm (12 inches) long and two cells thick that is encased in a rigid gelatinous sheath and resembles a lettuce leaf. Connecting the alga to its surface, the holdfast takes the shape of a disk. The life cycle alternates between generations that produce gametes (haploid) and spores (diploid). Accidental fragmentation is the most common mechanism for asexual reproduction.

Flour is a staple in many baking recipes and is made by finely grinding cereal grains or other starchy portions of plants. Baking items that need a light, airy texture are best made with wheat flour. Wheat flour is the most common type of flour in the West, hence the term "flour" is typically used interchangeably with "wheat" nowadays.

The starchy endosperm, or food-storage component, accounts for roughly 85% of a wheat grain's total mass, while the bran, consisting of many outer layers, accounts for about 13%, and the oily germ, or embryo plant, accounts for about 2%. The endosperm is isolated from the other parts of the kernel during the milling process used to produce refined flour. The entire wheat kernel is utilized in the production of whole wheat flour.

The wheat kernels used to make refined flours today are cleaned, tempered (by either adding or removing moisture), and cracked open by a set of rollers. Break flour is the refined and packaged form of the finest particles. The remaining bran and coarser endosperm particles (called semolina) are then sent through a series of rollers to be crushed into flour and the bran sorted off. To replicate the enhanced bread-making properties originally attained via natural age, bleaching and treating the flour is standard practice. The quality of flour is determined by the percentage of bran that remains after milling. (Choudhary, A. 2020)

When flour is combined with water to generate dough, the proteins in the flour are converted into gluten, an elastic component that forms a continuous network throughout the dough and is capable of storing gas, enabling the baked good to rise. The protein content of the flour is critical to the stability of the gluten. Flour made from soft wheat, which typically contains 8-12 percent protein, works well for soft, crumbly baked goods like cookies, pie crusts, and crackers. Bread, buns, hard rolls, and yeast-raised sweet rolls benefit from the higher protein content of flour made from hard wheat (about 12-15 percent).

Whole wheat, or graham, flour is made from the entire wheat kernel and is usually unbleached. Gluten flour is a starch-free, high-protein, whole wheat flour. All-purpose flour is refined (separated from bran and germ), bleached or unbleached, and suitable for any recipe not require special flour. Cake flour is refined and bleached, with a very fine texture. (Magdoff, F. 2012)

8.4 PULSES

Pulses are a type of leguminous crop produced mainly for the dried seed. The most well-known and eaten pulses are dry beans, lentils, and peas. (Pratap, A. 2011).

Green crops (e.g., green peas, and green beans) are not considered pulses; they are considered vegetable crops. Also omitted are crops produced primarily for oil extraction (e.g., soybeans and groundnuts) and leguminous crops farmed only for sowing (e.g. seeds of clover and alfalfa).

You most certainly consume more pulses than you know! All forms of dried beans, such as kidney beans, lima beans, butter beans, and broad beans, are popular pulses. Pulses include chickpeas, cowpeas, black-eyed peas, and pigeon peas, as well as all forms of lentils.

Pulses feature in staple meals and cultures all over the world, from hummus in the Mediterranean (chickpeas) to a classic full English breakfast (baked navy beans) to Indian dal.

Pulses are essential crops for a variety of reasons. They are high in protein and abundant in nutrients, making them a good source of protein, especially in locations where meat and dairy are not physically or monetarily accessible. Pulses are low in fat and high in soluble fiber, which can decrease cholesterol and aid with blood sugar management. Because of these properties, health organizations suggest them for the management of noncommunicable illnesses such as diabetes and heart problems. Pulses have also been demonstrated to assist in the battle against obesity. (Pratap, A. 2011)

Pulses are an essential crop for farmers since they may be sold as well as consumed by the farmers and their families. Farmers may preserve family food security and economic stability by having the option to consume and sell the pulses they cultivate. Furthermore, the nitrogen-fixing characteristics of pulses boost soil fertility, enhancing and extending farmland production. Farmers may improve farm biodiversity and soil biodiversity while keeping damaging pests and illnesses at bay by utilizing pulses for intercropping and cover crops (Magdoff, F. 2012).

Pulses can assist to reduce climate change by lowering dependency on synthetic fertilizers, which are used to artificially inject nitrogen into the soil. Greenhouse gases are emitted during the manufacturing and application of these fertilizers, and their excessive usage can be hazardous to the environment. However, pulses naturally fix atmospheric nitrogen in the soil and, in certain circumstances, liberate soil-bound phosphorus, minimizing the demand for synthetic fertilizers.

Pulse crops are members of the Leguminosae or Fabaceae families. Pulse crops are plant species that are collected primarily for their dry seed, which is used as both human food and animal feed. Field Pea, Lentil, Dry Bean, and Chickpea are the four most regularly produced pulse crops in Canada. On a smaller scale, Faba bean, Lupin, and Mung bean are also farmed in Canada. Soybean and Peanut are two more legume family members that most people are familiar with. These two, on the other hand, are not considered pulse crops, but rather leguminous oilseeds. It is usually considered that immigrants from various countries brought pulse crops to Canada (Pratap, A. 2011).

Because of their potential to create symbiotic partnerships with nitrogen-fixing Rhizobium bacteria in their taproot systems, pulse crops are considered ecologically beneficial. The nitrogen fixed in the soil is accessible to the subsequent crop. In addition, having pulse crops in the crop sequence minimizes disease incidence in crops that come after a pulse crop. A pulse crop's fruit is a pod containing one to several seeds that are abundant in protein, carbs, fiber, minerals, and vitamins (folate). Pulse intake as part of a regular diet may bring several health benefits, including a decreased risk of diabetes and heart disease.

More than 5.5 million t of pulses are farmed in Canada each year on roughly 2 million ha of land, mostly in five provinces: Ontario, Manitoba, Saskatchewan, Alberta, and Québec. Canada is a large exporter of pulse crops, with around 75% of yearly production exported. Canada contributes 35% of worldwide pulse commerce, with the pulse crop trade balance expanding from $863 million in 2000 to $2.1 billion in 2010. Commercial cultivation of pulse crops remains a successful endeavor for Canadian pulse producers. More than 10,000 Canadian farms cultivate pulse crops, with an annual farm cash receipt of around $1.7 billion. Pulse crop production has led to the rise of the pulse processing sector and the creation of job prospects in rural areas across Canada, in addition to being a profitable crop for producers.

To battle malnutrition in all of its manifestations, the globe needs to consume substantially more high-protein, low-fat, high-fiber pulse grains. The twin epidemic of under and over-nutrition causes a plethora of illnesses across the globe, including stunting, obesity, diabetes, and micronutrient deficiencies such as anemia. The cost to human health, productivity, and pleasure are immense. Better inclusion of chickpeas, lentils, beans, and other pulses in human diets can assist to reverse these negative tendencies. (Pratap, A. 2011).

Pulses also play a significant part in the agriculture industry, which is grappling with increased land and resource competition, climate change, expanding food demand, and complicated commodities markets. Pulses can boost total crop yield by increasing nitrogen availability, improving fertilizer and water usage efficiency, interrupting insect, weed, and disease cycles, and enhancing soil quality. Given geographical and natural resource limits for food production, pulse crops can feed a bigger population than animal protein can.

Increased pulse production and consumption take occur in the context of dynamic, interrelated global, regional, and national food systems. We need to understand how these processes function if we are to get more pulses onto farmers' fields and customers' plates (Magdoff, F. 2012).

Affordability, eating habits, taste preferences, and other cultural variables impact which socioeconomic groups consume pulses and their potential for expansion. While worldwide pulse demand has climbed in recent decades, per capita consumption has fallen. People consume 80% of pulses in underdeveloped nations; in rich ones, just 40% is consumed by humans and 50% is utilized for animal feed.

The bulk of pulses are eaten in the region where they are produced, and shortages in pulse production are rising in some locations. Because merchants control pricing, farmers in many locations do not get market signals promoting increasing output. Farmers will require better pricing signals, and consumers will need to be supplied with enticing pulse-based goods, to see large increases in pulse production and consumption. Pulses must be perceived as more than just a low-cost replacement for animal protein; they must be recognized as innately attractive and healthful. Pulse-based nutrition education has been proven to be successful in changing the dynamics of pulse intake and diet variety, and consequently child and mother health, in Ethiopia, India, Kenya, and Malawi.

Increases in pulse production, particularly at the domestic level, will be necessary to complement effective initiatives for persuading consumers to integrate more pulses into their diets. Despite limited funding, researchers have shown the potential to significantly improve smallholder pulse productivity and resilience, such as through the development of high-yielding, disease-resistant, waterlogging-tolerant fava bean and chickpea varieties in Ethiopia and biological control of pod borer in cowpea in West Africa. Diversification of pulse seed systems, along with broad marketing of recommended packages, has had a considerable influence in Ethiopia, which has one of the highest extension-to-farmer ratios in Africa. (Pratap, A. 2011).

However, greater output requires more than new pulse varieties and production techniques. Farmers' financial incentives, locally retained money from pulse production, and the impacts of subsidies and other agricultural policies all have crucial roles. Furthermore, societal conventions such as gender roles will determine who benefits from attempts to boost pulse production and consumption. Field findings from the Scaling-up Pulse Innovation project, for example, reveal that pulse agriculture, nutrition, and other social elements have been successfully integrated into food systems in two locations in southern Ethiopia. The initiative, which intends to reach 70,000 smallholder farmers, has brought healthy pulse-based snacks, pre-cooked flours, and porridges into remote farming homes, as well as school feeding and village shop models. To present, direct beneficiaries include roughly 15,000 female farmers and around 36,000 male farmers who got enhanced seed types of chickpea and common beans for planting.

Pulse value chains are quite diversified, ranging from transnational commodity export to local marketplaces offering traditional pulse varieties. Many value chains approach pulse crops as bulk products. Given fundamental commodity pricing, producers strive to maximize productivity and reduce production expenses. These value chains are vital to food and nutritional security, especially for traditional pulse-consuming communities and price is a significant factor. More and more, local and regional food businesses are creating specialized, value-added markets for pulse crops, especially if commercially viable applications can be discovered for the starch as well as the protein components of the pulse grain. Small and medium-sized food product manufacturers (e.g. baby food, breweries) and parallel industries (e.g. aquaculture) can incentivize farmers to meet specific quality parameters

There is a significant possibility for pulses to be promoted more prominently as ingredients in major food brands. Food firms have rising incentives to show the nutritional worth of their goods as well as the social and environmental sustainability of their supply chains. Increasing the proportion of pulses as ingredients in significant regional and worldwide brands can have tremendous influence. Consumers are likely to value food product attractiveness, convenience, and price (e.g. pulse protein fractions in ready-to-eat meals) Pulses are fundamental to many culinary traditions across the world and, in many countries, they represent a cornerstone of food and nutritional security. With the large growth in global food demands in the future, the role of pulses will become even more essential, especially in dietary protein and micronutrients. More focused evidence gathering is needed to clarify the importance of pulse consumption in reducing malnutrition and obesity and will provide data to inform national dietary guidelines and the emerging medical arena of 'prescription food' and to support policies that better incentivize farmers to grow pulses. (Magdoff, F. 2012)

8.5 FRUITS

According to botany, a fruit is a structure that develops from an ovary following a blooming period in flowering plants.

Angiosperms, or blooming plants, spread their seeds through their fruits. In reality, humans and many other animals have grown to depend on edible fruits as a source of food. Edible fruits, in particular, have long been spread by people and animals in a symbiotic connection that acts as a means of seed distribution for one group and nutrition for the other. Fruits, therefore, make up a considerable amount of the world's agricultural production, and some of them, like the apple and the pomegranate have acquired deep cultural and symbolic significance (Lloyd, B. 2012).

The term "fruit" is frequently used to describe the fleshy, seed-associated structures (or product) of plants that are sweet or sour and edible when they are still in their raw form, including strawberries, apples, bananas, grapes, lemons, and oranges. In botanical use, the word "fruit" also refers to a variety of structures that aren't often thought of as "fruits," such as nuts, bean pods, maize kernels, tomatoes, and wheat grains.

Many names used to describe fruits and seeds in everyday English don't match up with their botanical categories. A matured ovary or carpel that contains seeds, such as an apple, pomegranate, tomato, or pumpkin, is referred to as a fruit in botany. A seed is an ovule that has developed, whereas a nut is a sort of fruit.

A fruit is a particular plant's sweet or not-so-sweet (or sour) tasting product; nuts are a type of hard, fatty, non-sweet plant product that comes in shells (hazelnut, acorn). Typically, vegetables are savory or non-sweet foods (such as zucchini, lettuce, broccoli, and tomatoes), however, others can have a sweet flavor (sweet potato). (Lloyd, B. 2012)

Beans, peanuts, and peas are all legumes, as are cucumber, pumpkin, and squash (all cucurbits), as well as maize, eggplant, bell pepper (or sweet pepper), tomato, and beans. These botanically classified fruits are all popularly referred to as vegetables. According to botany, chili peppers and allspice are fruits. Rhubarb is sometimes referred to be a fruit when it is used in pies, yet the edible part of the plant is the petiole or leaf stem. Fruit names are commonly given to edible gymnosperm seeds, such as ginkgo nuts and pine nuts.

A type of fruit is a cereal grain, such as corn, rice, or wheat (termed a caryopsis). However, almost all of the edible grain fruit is a seed since the fruit wall is thin and attached to the seed coat. One or more blooms are fertilized, mature, and eventually turn into a fruit. The stigma-style ovary system is found in the gynoecium, which is in the center of the flowerhead and makes up all or a portion of the fruit. The ovary contains one or more ovules. Here, a female gametophyte generates an egg cell for the goal of fertilization, beginning the intricate process known as double fertilization. (A female gametophyte is referred to as a megagametophyte, and it is also referred to as the embryo sac.) Ovules will develop into seeds following double fertilization. (Lloyd, B. 2012)

The first stage in fertilization is pollination, which is the transfer of pollen from the stamens to the stigma-style ovary system inside the flower head (see graphic). After pollination, a pollen tube develops from the pollen (deposited) down the stigma into the ovary to the ovule. Two sperm are delivered to a megagametophyte from the pollen. The second sperm enters the central cell to produce the endosperm mother cell, completing the double fertilization process. One sperm combines with the egg to form a zygote inside the megagametophyte. The endosperm mother cell will later grow into endosperm, a nutritional component utilized by the embryo, and the zygote will later develop into the seed embryo.

As the ovules transform into seeds, the ovary begins to ripen. The ovary wall, called the pericarp, may become mushy (as in berries or drupes), or it may produce a hard outer coating (as in nuts). In some multi-seeded fruits, the amount of fleshy structure development is inversely correlated with the quantity of fertilized ovules. The exocarp (outside layer, also known as epicarp), mesocarp (middle layer), and endocarp (inner layer) are the two or three separate layers that make up the pericarp.

Some fruits lose their sepals, petals, stamens, and/or flower style as the meaty fruit ripens. For simple fruits made from inferior ovaries or those that are attached below the other floral components, there are certain portions (such as petals, sepals, and stamens) that merge with the ovary and mature with it (see graphic reinsertion point'). An accessory fruit forms when floral components other than the ovary make up a substantial portion of the fruit. Examples of accessory fruits are the apple, rose hip, strawberry, and pineapple (Lloyd, B. 2012).

It is essential to examine the floral structure to comprehend how a particular fruit arises since numerous elements of the flower, besides the ovary, can influence the construction of a fruit. Fruit development may be divided into three categories in general:

Simple fruits called apocarpous fruits to grow from a single bloom (while having one or more separate, unfused carpels).

- A single gynoecium produces syncarpous fruits (having two or more carpels fused).
- A flowering plant's inflorescence develops numerous fruits from numerous blossoms.

There are 8 main types of fruits. They are:

- Simple Fruit: These fruits develop from the ovary of a single flower when it reaches maturity. There are two categories for simple fruits: dry and fleshy. Both dehiscent and indehiscent dry fruits can be found. Brazil nuts, strawberries, coconut, walnuts, hazelnuts, and other dry simple fruits are examples. When fully developed, fleshy simple fruits have a partially or completely fleshy pericarp (fruit wall). Some examples are cherry, tomato, olives, peach, plum, and cranberries.
- Aggregate Fruit: These fruits develop from a single flower's ovaries. The four main types of aggregation fruits are fruits in the shape of berries, drupelets, follicles, and achenes. Raspberries, blackberries, and custard apples are a few examples.
- Multiple Fruit: A flowering plant's inflorescence, or flower cluster, bears several fruits. These flowers each produce fruit on their own, and when the fruits are fully developed, they come together to form a mass. The list of fruits includes mulberries, pineapple, figs, and others.
- Berries: Since it develops from a single ovary of a single flower, a berry is a straightforward fruit. The two most common berry varieties are blueberry and cranberry.
- Accessory Fruit: Any fruit whose edible portion is formed from the ovary as well as other components of the flower is considered an accessory fruit. It might be a single fruit, a group of fruits, or a combination of fruits. Some of the most well-known examples include strawberries, pineapple, apples, and stone fruit.
- Seedless Fruit: There are two ways that seedless fruits can grow. The first occurs when parthenocarpy is absent from the fruit, whereas the second happens when pollination encourages fruit development. They are the most prized fruits since the majority of people eat them.

Young people's emotional health may be improved by eating more fruits and vegetables. According to studies, a good mood may make people choose healthy diets over-indulgent ones. The benefits of fruits are described in more depth below;

- Boost Energy: When you consume fruits, your energy supply quickly increases; this is one of the primary benefits of fruits that we may use in our hectic schedules. This is why athletes frequently consume fruit during and after exercise, and why pregnant women's diets nearly always include fruits.
- Heart Health: Fruits like the apricot, apple, banana, cantaloupe, berries, grapefruit, and orange are good heart defenders since they are loaded with flavonoids, carotenoids, fiber, potassium, and magnesium. Vitamins including vitamin A, vitamin B6, vitamin C, vitamin E, vitamin K, and folate are also found in fruits. These vitamins all help to regulate cholesterol and avoid diseases like stroke, atherosclerosis, and heart attacks.
- Diabetes: The majority of people globally are afflicted with type 2 diabetes. Although all fruits are healthy, diabetics should avoid those with a high carbohydrate content. Fruits like apples, avocados, cherries, bananas, oranges, peaches, and plums that have a low glycemic index (GI) of less than 55 help to control blood sugar levels. Consume fresh fruits to get the maximum benefits; processed and canned fruits are bad since they contain artificial sweeteners.
- Cancer: Many malignancies, including liver and breast cancer, can be prevented and treated with the use of fruits rich in vitamins. Examples of these include citrus fruits like oranges and tangerines, goji berries, Camu Camu, and soursop. When consumed often, fruits can exhibit hepatoprotective properties that help prevent cancer.
- Blood Pressure: Bananas, apples, melons, plums, pears, apricots, and mangoes are a few examples of fruits strong in potassium that might help lower blood pressure. This mineral's vasodilating properties have been connected to its ability to control blood pressure.
- Prevent Diseases: Because they include powerful flavonoids, antioxidants, minerals, vitamins, phytochemicals, and a variety of micro and macronutrients, fruits are very good for your health.

Regularly eating fresh fruits lowers the risk of developing diabetes, heart disease, cancer, high blood pressure, stroke, indigestion, and other chronic disorders. Some fruits, like bananas, include vital elements like potassium, which helps to lower blood pressure, reduce anxiety, and prevent strokes. Consuming fruit almost removes vitamin and mineral deficiencies and the symptoms they cause. Fruits are also rich in water and fiber, which support healthy digestion and maintain a healthy weight.

- Aid in Digestion: Fruits that are fibrous aid with digestion. Additionally, rich in dietary fiber, the skins of several fruits help with proper digestion and elimination while guarding against conditions like constipation and gastritis. However, the skin is thick and cannot be consumed by people in other instances, such as lemons, bananas, melons, and oranges. In certain cases, the edible portion of the skin's meat is also quite high in fiber.

  Fruit's high fiber content not only provides a wonderful laxative effect but also fills you up by supplying bulk nutrients. Fibrous fruits also reduce hyperlipidemia, hypertension, diabetes, and obesity, which all contribute to the development of heart disease. Fruits are known to prevent colon cancer and combat bowel irregularity illnesses because of their fibrous nature and anti-carcinogenic properties.
- Weight Loss: You'll notice that your weight is being controlled and that your health is becoming better once you start eating more fruits. Fruits, however, cannot help you lose weight on their own. Regular exercise is also a good idea.

Include fruits in your family's diet to lose large amounts of weight, according to several American research findings. Dr. Joanne L. Slavin of the Department of Food Science and Nutrition at the University of Minnesota claims that a person who frequently eats fruits would be less prone to snack on junk food and unhealthy foods in between meals. Fruits also contain 90–95 percent water, which has a powerful diuretic effect on the body, flushing out unwanted toxins and nitrogenous wastes.

A great example is the fruit lychee, which has very few calories, no cholesterol, and is a superb laxative. Additionally, it helps with body weight loss. Fruits are normally fat-free, except for coconuts, avocados, and olives. (Magdoff, F. 2012)

- Skin Care: Fruits enrich your skin with essential vitamins, minerals, and antioxidants, keeping it supple and moisturized while also preserving your youthful skin for an extended period. If you're tired of using over-the-counter acne creams, think about including berries, apples, or bananas into your daily diet and watch the acne clear up quickly. One of the most annoying skin issues that may affect people of all ages is acne. Instead of depending on expensive dermatological treatments, you may treat this issue simply by eating enough fruit. Although there are additional causes of acne, skin infections and dermatological conditions are the main ones. Apples are the best alternative for treating acne; consume the apple skin, which has a high amount of pectin and prevents constipation, a condition that can aggravate acne. Bananas' high fiber content, which lessens constipation, also helps to minimize acne. Berries have amazing cleansing properties and shield you from skin issues. When consumed during the day, grapes help your skin and body cleanse. While papayas rejuvenate your skin and repair little damage that you might not even notice, lemon juice gradually fades the signs and scars of acne spots.

Aim to be careful of your other eating habits in addition to including fruits in your diet. You should consume less caffeine since it increases certain hormones in your body that lead to acne. Reduce your intake of dairy products and red meat as well since these foods might worsen acne because they increase your body's production of hormones. Additionally, you should keep your skin clean by washing it frequently with herbal soap or a mild face wash.

- Healthy Hair: Fruits greatly contribute to long, healthy hair in addition to improving the appearance of your skin. Your hair's texture is softer and gains luster thanks to vitamin A. Fruit consumption on an empty stomach can help stop hair loss and early graying. Fruits with citrus peels are excellent hair conditioners. Citrus fruits contain vitamin C, which maintains your hair healthy and shiny while preventing hair loss. It is also believed that kiwis, strawberries, and pineapple encourage the growth of hair. Fruits may also encourage the growth of healthy hair and maintain the luster and silkiness of your hair.

Fruit benefits for healthy hair can also be gained by using fruit-infused hair masks. Melon and banana hair masks are highly recommended by several beauty experts for at-home hair care. Additionally, you might use an aloe vera and lemon juice paste on your hair. Before completely washing your hair with a mild shampoo, let the paste sit on your hair for 45 minutes. These packets not only improve hair quality but also soothe itchy and dandruff-prone scalps. Avocado has a sizable amount of vitamin E, which is known to be good for hair. Avocado hair masks give your hair more power and stop it from going gray too soon. Coconuts are also great for growing hair. Tropical nations frequently treat their hair with coconut oil. (Magdoff, F. 2012)

Chapter

9

MODERN FOOD PROCESSING TECHNOLOGIES

Contents

9.1 INTRODUCTION

Only approximately a third of the roughly 5,000 food products in a typical store is fresh. The remainder has undergone some processing, such as canning, waxing, drying, freezing, bottling, pickling, packaging, wrapping, baking, and so on. One of the key purposes of processing is to increase food shelf life. Bacteria is frequently the first thing that comes to mind when considering food degradation. Because acidic fruit is a favored meal for yeasts and molds, it will turn sour, yeasty, or moldy over time. The introduction of bacteria is responsible for most food quality reductions and purifications. Regardless of its existence, other chemical changes occur before, during, and after processing. Enzymes are found in every cell and help to speed up numerous chemical processes. For example, when food is canned, the heat in the process eliminates the bacteria and yeasts that cause rotting. To avoid future contamination, the cans are hermetically sealed. Dehydrating and freezing are means of preservation, yet both deprive organisms of the water and warmth they require to flourish. Some animals will be slaughtered by any method. The organisms' growth rates revert to normal if water or heat is given. (Barba, F. J. 2017).

Storing foods at a cold temperature, sheltering them from light, adding substances that hinder chemical reactions, and minimizing the length of time the food is in contact with air are all common strategies for slowing down chemical changes. Because all B vitamins, vitamin G, and other minerals are dissolved in water and might be lost during processing, employees must be aware of this. Fatty acids, such as the oil found in canned fish, degrade vitamins A and D. Vitamins A, C, and D degrade from air exposure. Heat, even in the absence of oxygen, destroys vitamin B in non-acidic foods. Exposure to light destroys all vitamins B, A, and G. Warm meals deplete these nutrients more than cold meals. The quality of a final product is initially restricted by the raw material utilized to manufacture it.

A meal will turn out differently than you want it to if it contains rapidly decaying fat, starchy peas, green fruit, or a strawberry kind with little flavor or color. Not all raw materials are suitable for processing in the same way that not all apples are edible, and not all potatoes are suitable for baking. Fresh market potatoes are accessible in almost every state, but only certain types grown in specific places are suitable for drying and freezing. Many procedures will be required on the same object. The use of potatoes with a low water content benefits dry mashed potatoes. The waxy variety is required for canning. Mealy potatoes float in salt brines while other varieties sink, allowing them to be distinguished. When cost is an issue, it may be essential to adopt alternatives to the gold standard. It may be necessary to include less appealing types in the pack within a given year to extend the processing season past the narrow window in which the most favored kind is available. After deciding which components to utilize, the processing company must make arrangements for their timely and icy delivery. Until recently, consumers had to wait long periods for fish and shellfish. When a large quantity of fish is captured, it is promptly frozen and packed in crushed ice. Many are preserved by freezing onboard ships and dissecting them once they have thawed on shore. Some food on board ships has been processed (canning or freezing). This helps to preserve the flavor and quality.

Whether fresh or processed, milk, eggs, meat, and poultry require immediate cooling and refrigeration. Peas, lima beans, asparagus, and sweet corn are canned vegetables that degrade quickly. If they are not processed or chilled soon, flavor, sweetness, and overall perfection, as well as vitamins B2 and C, are swiftly lost. Peas provide a unique problem since some of their vines develop in unexpected locations far from the main plant. When exposed to heat or damaged, peas quickly lose their flavor. Spraying water on the vines or peas at the plant's vining stations to keep them cool is now usual practice. The first processes in the processing facility include cleaning, sorting, grading, and preparing the food for consumption. (Barba, F. J. 2017).Deboning the chicken, peeling the fruit, and squeezing the milk are all examples of this. Food safety and cleanliness must be prioritized in the food sector. Inspecting a factory is as simple as observing how clean it is. There appears to be a substantial amount of dirt that has to be cleaned. Packaging uniformity is critical for sorting and grading. The extent to which a processor may divide materials into sub lots tailored for certain usage is determined by the breadth of his product line. Some individuals believe that differences in how things look have been overemphasized, but they do not believe that differences in how things taste have been overemphasized.

The food is transported through a flume, a trough filled with water. Many nutrients are lost in the process, including vitamin C, B vitamins, and minerals. Conveyor belts have taken the place of flumes in several industrial operations. Most foods are processed after they have been prepared for human consumption. Blanching, pasteurization, concentration, or sulfation are among the examples. Pasteurization increases the shelf life of liquids such as milk, cream, eggs, and fruit juices. Pasteurization is a moderate heating procedure that kills all or almost all germs, depending on the substrate. Because enzymes are inactivated, food flavor and color can be preserved. However, after being pasteurized, condensed, and frozen orange juice loses part of its fresh flavor and ability to separate into clear and pulp layers. (Barba, F. J. 2017)

9.2 PROCESSES OF FOOD PROCESSING

During batch pasteurization, liquids are generally heated to roughly 165 degrees Fahrenheit. These treatments, which used to take 15 or 30 minutes, may now be done in a couple of minutes by employing a fluid that is rapidly heated and cooled as it goes through a series of coiled tubes. Color and taste might be enhanced in nanoseconds. Before being sold, most fresh food is blanched. Blanching is the process of quickly cooking food in water or steam. Peas and other comparable tiny objects are cooked in boiling water for roughly a minute or longer at lower temperatures. Larger volumes need more cooking time. During the blanching process, enzymes in food are eliminated, allowing for greater color and flavor retention. Unfortunately, there is no widely accepted method of blanching at the moment. Some processing facilities can cause water-soluble vitamins and other components to lose up to 10% of their weight. Some folks have lost 40-50% of their body weight with just one meal. Long-term low-temperature treatment results in a greater loss than short-term high-temperature treatment, and water blanching results in a greater loss than steam blanching. Because of the great color enhancement, it delivers, rapid high-temperature blanching has supplanted lower-temperature techniques for processing most frozen green beans. Electronic blanching has been explored intensively since 1959 to reduce leaching losses, but no practical technology has yet been identified. Instead of blanching, sulfite processing is sometimes used on dried or frozen fruits and vegetables. After blanching, some have been suited. Sulfating can occur through the emission of sulfur dioxide gas or the immersion of the meal in a sulfur dioxide-containing solution. Sulfite prevents food color and flavor loss by binding to enzymes, thereby reducing their action. If the chemical is stored at the same temperatures as dried food, it will degrade with time. After a while, the shield will be removed. Despite being harmful to vitamin B, sulfite paradoxically aids in preserving vitamin C. The overwhelming sulfite flavor turns off most people's taste buds. We never blanch dessert ingredients like berries, peaches, apricots, and cherries because we cherish their original flavor. Instead, a sugar syrup coating is applied to keep the product fresh and prevent oxidation. To extend the shelf life of apricot, peach, and cherry syrup and keep it from spoiling, vitamin C is occasionally added. Instead of the natural components of the fruit, oxygen clings to the extra vitamin.

The efficacy of vitamins may diminish with storage. It is standard practice to partially concentrate soups, milk, fruit juice, and other liquids before preserving and freezing them for practical reasons. A vacuum is required to remove water at a lower temperature. Using this procedure, the food retains practically all of its high-quality and nutritional value. Concentration is employed in an unusual method here to extract the essence of the fruit. It preserves the flavor's building elements. The fragrant vapors emitted by jelly and other fruit-based delights during their first cooking phases are collected, condensed, and concentrated. The essence works nicely in sweets such as ice cream and frozen yogurt and beverages such as juice and jelly that have a fruity flavor. Radiation as a method of food preservation is a viable option. Even though no irradiated food could be legally sold in 1958, millions of dollars have been spent on research into irradiation. Gamma rays or high-speed electrons can destroy microbes and insects in food in the same way as X-rays can kill cancer cells in the human body. Irradiated sterilized products are safe to eat. Otto Wüst invented irradiated food in 1930, but it wasn't until 1947 that real research was performed.

Low doses of radiation were employed in tests to postpone fruit ripening, eradicate insects in dry commodities such as flour and spices, and prevent potatoes from sprouting. Minor levels like this normally have little effect on the flavor or presentation of food. Radiation levels in food are almost always provided in rads nowadays. The quantity of energy emitted by one gram of radioactive material is measured in rads (100 ergs). A sort of "pasteurization" happens when subjected to radiation at a medium intensity of 100,000 to 1,000,000 rads; practically all bacteria are destroyed, but the resulting product is not fully sterile. Fruits and vegetables have a longer shelf life after being exposed to this temperature range, but meats and fish have a longer shelf life after being refrigerated. While 1 million rads produce a considerable quality loss in many foods, it is sometimes possible to discover a smaller number that may improve quality retention without significantly compromising flavor, color, or texture. This level of radiation supplements typical preservation procedures such as canning, freezing and drying. Irradiated products, including canned foods, may be preserved for a long period if properly packaged and stored. Concentrations this high, however, have a disagreeable flavor, particularly in meat and fat, and cause vegetables to become notably mushy. Researchers strive to minimize the process's expensive costs and achieve consistency without losing food quality. In

1959, however, studies had not yet established that irradiated food was safe. Food items frequently include several additives. Colorants, flavoring ingredients, mold inhibitors such as propionates in bread, antioxidants that delay one type of fat rancidity, emulsifiers such as those used to prevent separation in French dressing, flour bleaching agents, thickening agents, and ingredients that prevent powders from caking are just a few examples. These supplementary materials are referred to as additives.

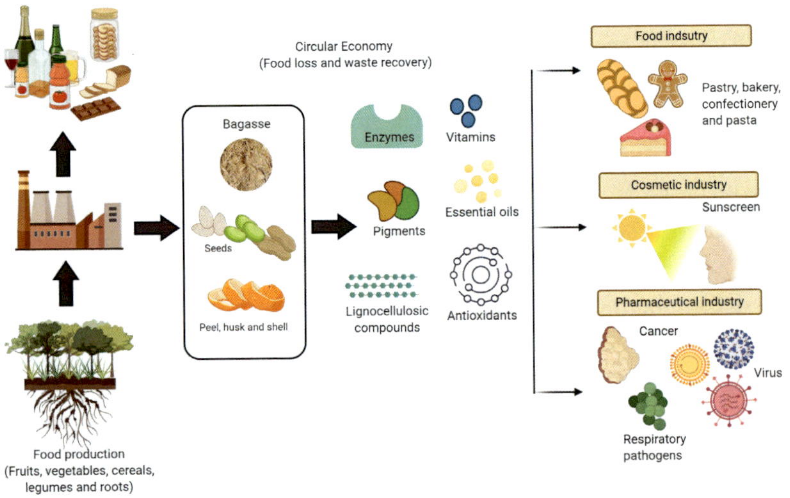

Figure 9.1: Development of a layout model suitable for the food processing industry.

Source: Osorio LLDR, Flórez-López E, Grande-Tovar CD. The Potential of Selected Agri-Food Loss and Waste to Contribute to a Circular Economy: Applications in the Food, Cosmetic and Pharmaceutical Industries. *Molecules*. 2021; 26(2):515. https://doi.org/10.3390/molecules26020515

In 1938, the number of additives utilized was rather low. That year saw the passage of the Federal Food, Drug, and Cosmetic Act, which protects consumers' health. However, scientists and engineers quickly discovered that chemicals might aid food processing in various ways. Several compounds that were either unknown or underused in 1938 are today widely used. Concerns about the possible hazards of some chemicals began to circulate among the general population. In 1958, Congress enacted a new version of the Federal Food, Drug, and Cosmetic Act after considerable thought and inquiry. Discuss these developments in further depth. The 1938 restrictions made determining the safety of food additives in advance difficult. It was up to the Food and Drug Administration to identify them in goods currently on store shelves and conduct testing to "prove they are hazardous or harmful" to the court's satisfaction to have them removed from sale. The FDA was overwhelmed by the number of additions, and those tested may still be sold pending the results of more rigorous research. (Barba, F. J. 2017).

These concepts are applied throughout the aseptic canning process. The dish is prepared to absorb heat quickly. After that, it is sealed in aseptic cans to prevent contamination while being processed. In 1959, the breakthrough process was used to make milk, fruit juice, and creamed soup. The lab has successfully produced baby food, maize, peas, chopped vegetables, tomato juice, and various fruit liquids. The novel method should work well for storing any liquid, purée, or food in microscopic pieces. We'll have to wait until technology like high-frequency electronic heating becomes more common before eating meat and fish. Adding chemicals that boost the heat's sporicidal action on bacterial spores is one unique technique for decreasing canning damage caused by heat. Because of the chemicals, less heat is required to kill the germs than traditional canning. 26 of the 650 compounds studied before 1958 were shown to enhance spore death; however, many of these chemicals are poisonous or have unpleasant odors, rendering them unsuitable for use in food. Subtilin and nisin are two of the most promising antibiotics. Several varieties of cheese naturally contain nisin. Powerful compounds include propylene oxide, diepoxybutane, dodecylguanidine, and hydrogen peroxide. Before they may be used on commercially canned food, it must be proved that they effectively remove the spores introduced into experimentally made food. Several canning trials have shown encouraging outcomes. Animal feeding tests must demonstrate that any chemical for commercial use is safe for long-term exposure. The testing process might last up to two years. Dehydrated nutrients are often sold at cheap prices. For the great majority, little preliminary planning is required. They're lightweight and tiny, making them easy to store. They should be deployed quickly in the case of a transportation breakdown. Many individuals have worked hard to enhance the flavor and texture of dry meals.

Raw materials that are more amenable to dehydration processes have been used; new processing technology has been implemented; established processing techniques have been meticulously applied; sulfiting for vegetables has been reduced; equipment, particularly dryers, has been improved; finished product moisture content has been reduced, and inert gas packaging has been implemented. Dehydration is a way of preserving food that eliminates most water content (up to 99 percent in some situations) without affecting its flavor or texture.

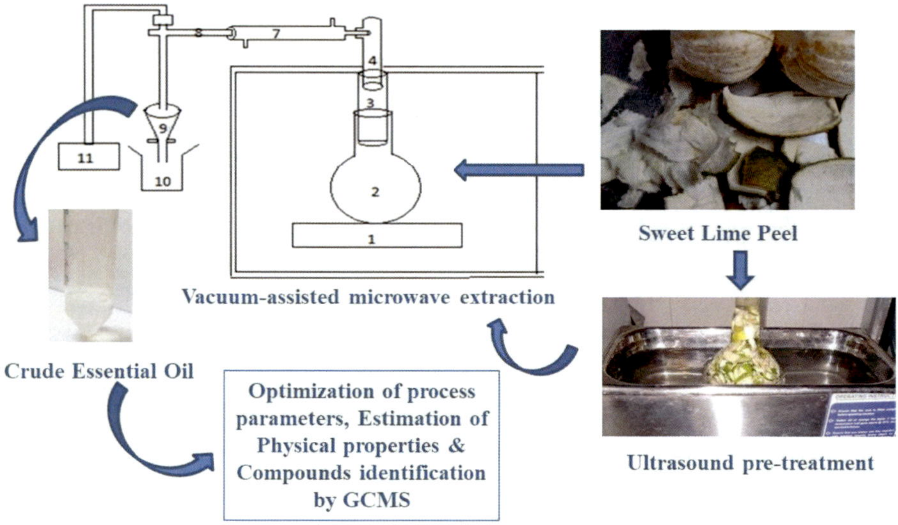

Figure 9.2: Lime processing.

Source: Arafat Y, Altemimi A, Ibrahim SA, Badwaik LS. Valorization of Sweet Lime Peel for the Extraction of Essential Oil by Solvent Free Microwave Extraction Enhanced with Ultrasound Pretreatment. *Molecules*. 2020; 25(18):4072. https://doi.org/10.3390/molecules25184072

Heat results in a brownish tint, a scorched flavor, and a distinct texture. The first half of the water is swiftly removed with minimum quality loss. After practically all of the water has been removed from the meal, most of the harm happens. As spray drying technology improves, less heat damage occurs throughout the drying process, resulting in higher-quality dried eggs. Before drying, the egg mixture was given a beneficial treatment. An unpleasant chemical formed when the egg's inherent sugar and a protein component react ensures the decomposition of dried eggs. The sugar in the new method has been modified to not react with the protein. Milk drying times have decreased as spray drier technology has improved. On the other hand, the new skim milk powder particles have a sponge-like structure due to a second drying cycle, which explains why they rehydrate so rapidly. Powdered, quick skim milk made its debut in 1954 after 30 years of study and development. Before the rest of the material is dry enough to be stored in a forced-draft tunnel, heat usually destroys part of it. The dryer's performance is not as dependable or high-quality as that of certain newer dryers. The fine food is driven into the heated air of the fluidized-bed drier, where it travels like a liquid and dries in the air. This device is used in the business sector during the granulating phase of the final drying process for dried mashed potatoes. (Barba, F. J. 2017).

The food is evenly dried by tumbling it on the belt of the unique belt-trough continuous drier. Another dryer keeps the food warm by continually dispersing it on its conveyor belts. If the vacuum is powerful enough, it can dehydrate anything at a low drying temperature and without air. The raw, original product's quality will be unaffected by drying. Under a high vacuum, foods may be practically completely dried, allowing for easier packing and longer storage. Because of their high sugar content, which causes caking and stickiness, fruit juices can be difficult to dry using typical techniques. The high-vacuum method, on the other hand, is effective. Furthermore, many fruit liquids, particularly citrus juice, quickly develop a disagreeable flavor when exposed to air. The strategy was so effective in 1958 that it significantly increased orange and grapefruit juice powder sales. Powdered tomato juice and orange-grapefruit juice, which may be reconstituted into single-strength juice, puree, or paste, were set to enter the market soon. For example, tomato powder and dry components can be used to make spaghetti sauce and tomato aspic. Vacuum drying can be used for batch drying on shelves and continuous belt drying of liquids, eggs, whole or juiced fruits, and vegetables. Because the treatment was rather costly in 1959, it was not frequently adopted. Freeze-drying, another low-temperature drying method, has demonstrated encouraging results, particularly with meats, which are notoriously difficult to dry.

9.3 PACKAGING OF PROCESSED FOODS

Dating dehydrated items in the package is frequently required to remove any leftover moisture. The delivery concludes with a little envelope holding a chemical. Food absorbs a lot of moisture when it's in storage. While more expensive, the treatment minimizes the quality degradation in the later stages of natural dehydration. During the International Geophysical Year, Antarctic explorers ate various dry meals, ranging from normal rations to experimental packs produced using cutting-edge technology. Two dried beefsteaks, dehydrated onions, mashed potatoes, green beans, crackers, canned butter, canned fruit cake, and dry, instant chocolate milk are typical pound-weight supper ingredients. Even though grains aren't often thought of as dried meals, their dryness is the fundamental reason they survive so long. Wheat is dried in the field, either partially or completely. Before buying cereal, customers should read the ingredient list to determine if the whole seed or the endosperm is contained (the interior of the kernel). The seed's outer bran layer and oily germ are nutritionally vital but are often overlooked due to their unpalatability or poor storage qualities. White flour is made by grinding the seed's endosperm into a fine powder, mixing many batches to get the required characteristics, and occasionally employing chemicals to speed up the wheat's natural bleaching and maturation. Separating flour into smaller volumes for specific usage is a recent invention in the milling business. Air currents of varying speeds keep particles of varying densities apart. Bakeries and dry baking mix producers frequently acquire flour that fulfills the functional parameters of a certain recipe.

As a result, a home baker is more likely to get excellent and consistent outcomes with homemade and store-bought baked goods. Nutrients are added to cereal meals to compensate for those lost during processing. Wheat flour combines vitamins B2, iron, niacin, and thiamine. Many popular foods have been made healthier. The National Research Council of the National Academy of Sciences has approved the fortification of flour, bread, cornmeal, and white rice. The Food and Drug Administration has allowed a certain amount of each of the several bread and wheat additives. There are two commercial ways and many do-it-yourself approaches for enhancing the protein content of breakfast cereals. Blends of different cereals, such as wheat, oats, barley, and even flaxseed, are occasionally used to get a higher protein content. Extra proteins, such as wheat gluten, soy protein, or milk casein, are occasionally included. There are several sorts of ready-to-eat morning cereals on the market nowadays. Some of them have sugar added to them. Pre-packaged cereals with the same nutritional content as handmade cereals are usually costlier. Cereals and cereal-based meals provide more than 20% of our daily calories and virtually all of our vitamin B. We stock up on various dry baking ingredients to produce quick-baked delights without using white flour. They are made scientifically and regularly produce good results when completed at home. Baking mixes come in a range of flavors, including those for cookies, pie crusts, buns, muffins, angel cakes, butter cakes, and fruit cakes.

Hoppers, feeders, and conveyors are commonly employed in plants to mix dry components mechanically. The oil is then added to the mixture by spraying it over the dry components or physically cutting it in. The development of rancidity in the mixtures has been a major source of worry.

It ruins the flavor and texture of baked goods and imparts a disagreeable flavor. Another source of worry is the leavening agents. Pre-baking reactions may diminish the volume of the finished product. Both concerns have been simplified by drying the flour to a water content of 6-7% rather than the standard 13-14%. If the baked product had a strong baking powder flavor, the baker was concerned that there would not be enough leavening agents. The dry soups are different dried component combinations distributed in cartons. Each component goes through its drying process. Belt-trough dryers are used to dehydrate vegetables, drum dryers for beef extract, and tunnel trays for dehydrating noodles. To finish, combine the dry seasonings and spices with the remaining ingredients. The completed soup isn't much better than its constituent elements. One example of how designers constantly search for ways to improve everyday items is an instant soup that rehydrates in hot water.

In 1955, the edible weight of frozen food output increased from 275 million pounds in 1932 to 7,329 million pounds. Because some goods are not sold in bulk, their prices are not included in the estimates. When these variables were included, total production in 1955 was roughly 10 billion pounds. According to current commercial procedures, freezing has little to no influence on foods' flavor, color, or nutritional value. Suppose a frozen meal fails in these areas. In that case, we frequently discover that it was made with defective raw materials, was improperly processed before freezing, or was maintained at an abnormally high temperature for an extended period. Freezing is rarely a problem, but if food freezes too slowly, it may lose some flavor, color, and nutritional value. Freeze-dried food has an odd texture. A few textures can only be changed by freezing them. Lettuce and tomatoes, two raw salad mainstays, lose vitality and become worthless. Typically, fruit is ripened to a softer texture. Even without adding salt, sugar, or glycerin, egg yolk turns rigid and loses mobility when thawed. However, egg white keeps its fluidity. Because of the presence of these compounds, the shelf life of frozen eggs is limited. The packaging of frozen goods has undergone several changes.

Many frozen goods would last significantly longer in airtight, hermetic containers than paperboard or other non-hermetic materials. Few things are packaged this way due to the rising cost of airtight containers. Skin darkening is caused by dehydration, most visible in chickens. The impact might be minimized by covering the birds with impermeable materials like polyethylene. Dehydrofreezed food keeps all of the benefits of both drying and freezing. Dried fruits and vegetables lose weight and volume throughout the drying process, but their quality may also suffer. While freezing has advantages, storage and shipping costs can pile up. Dehydrating fruits and vegetables to around half their original weight and volume protects their quality using the dehydrofreezing preservation procedure. Dehydrofrozen food has the same nutritional content as frozen produce but costs a fraction of the price to package, freeze, store, and transport internationally. Dehydrofrozen produce includes peas, carrots, potatoes, apricots, cherries, boysenberries, and apples. The housewife now only purchases recipes that have been remanufactured using dehydrofrozen components, such as fruit and vegetable pies and pastry-based vegetable soups.

Since the 1940s, most developments in the food sector have favored frozen and cooked choices. There are presently 200 distinct versions of these meals, with a total cost of one billion pounds. Soups, meat, poultry, fish, exotic food, baked goods, desserts, and entire dinners are just a few examples of what you may make at home. Many ready-to-eat dishes may be made using frozen ingredients and fresh ones. Most goods are bread, rolls, cakes, pies, pastries, and waffles. They keep roughly the same quality over long storage periods as they did immediately after freezing. They require coverings to prevent thawing moisture from touching the food surfaces in the packaging. Several items' stability is called into doubt. The meal may sometimes be fresher for longer by covering it with a sauce or gravy rather than having air between the food and the container. A lack of oxygen, which delays the oxidation process, can generate stale and other undesirable tastes in pre-cooked meals. White sauces, puddings, cream pun fillings, eggs, and other common starchy thickeners produce separation or curdling in frozen gravies. Curdling can be avoided by using thickening agents and stabilizers that are not readily available to home cooks but are utilized in industrial kitchens. Because of their lack of hygiene, frozen foods should be treated as perishable from when they are manufactured until consumed. (Barba, F. J. 2017).

While certain germs and viruses can be eliminated by cooking, others are constantly present and must be handled. They begin to form when the dish defrosts. When food is stored for an extended period after it has been defrosted, unpleasant scents and smells arise, the consistency and look change, and the food eventually degrades. Many frozen foods deteriorate faster than their fresh counterparts, therefore they should be used as soon as possible after defrosting. The first test identifies the overall number of organisms in the meal, while the second assesses whether or not the food contains pathogenic organisms that can make people sick. Even under the most sterile circumstances, frozen meals may contain some microorganisms. Foods with a low bacterial content reduce the chance of being ill from eating anything. Furthermore, meals made in a low bacterial concentration environment are likely to be more nutrient-dense and of higher quality. Although we prefer frozen meals with a low overall bacterial count, the presence of potentially harmful microorganisms is more concerning. Bacteria can be found in the soil in which crops are grown, the hands of employees in food processing plants, and animals' meat and other items they consume. Most food safety experts believe frozen meals cooked in the same way as fresh ones are just as safe to consume. These goods include frozen vegetables and uncooked meats. Both are cooked, not just reheated, before serving. The conditions that apply to ready-made frozen foods are vastly different. They provide a range of unique issues that necessitate improved security measures throughout the processing facility and up to the point of consumption.

Precooked foods are more likely to be contaminated before freezing than frozen entrees. Most of them include many substances, some of which may be detrimental to the environment. Manual dexterity is required for tasks such as deboning chickens by hand. When many items for one package are prepared independently, delays in various preparations may result in a rise in germs. The French fries for a plated meal may be finished before the gravy. Because certain pre-prepared foods are not thoroughly cooked like their homemade counterparts, more vigilance is required while ingesting them. Despite being branded "ready to eat," many of these frozen meals are not prepared at home. They have only received the heat of reheating. To avoid the formation of germs, these items must be frozen until ready to cook. There have been very few reported incidents of disease linked to frozen meals. However, new processes and items are frequently introduced into the market, needing extraordinary vigilance to ensure health commodities' production. (Chemat, F. 2021).

Food manufacturers and distributors share responsibility for guaranteeing the safety of their products with homemakers and regulatory organizations. It all began in 1956 when high-ranking leaders from the food and pharmaceutical sectors and government organizations launched an enormously significant project. This initiative aims to establish safe methods for the retail distribution and sale of frozen meals. The voluntary code would serve as a guide for various state and local food and drug enforcement agencies. The code for frozen ready-to-eat meals was scheduled to be completed in 1959. No matter how thoroughly we inspect our items at each stage of manufacture, we can never guarantee a high-quality supper. Quality may be modified at nearly every stage of the manufacturing process. As a result, store-bought processed food can't compete with farm-fresh fruits and vegetables, wild-caught fish, and handcrafted sweets. We are content with manufactured goods when we cannot afford farm-fresh ingredients and do not have time to bake a cake. Let's start with the total cost of food and then look at how other things influence that cost. In 1941, the United States spent up to $20 billion on food. In 1958, trade totaled almost $70 billion. In 1958, they spent around a quarter of their cash salary on food. If people spent the same amount of their income on food in 1958 as they did between 1935 and 1939, the national average family budget would have been $16,400. Our cuisine is undeniably more expensive. This rise results from several reasons, not just rising processing costs. Several additional variables play a role, including an increase in the frequency and relative cost of eating out and a general trend toward more premium purchases and marketing margins.

The Census of Manufactures contains figures for the total processing cost. In 1939, production increased the value of food and related commodities by $3.5 billion. This figure was $13.5 billion in 1954. After adjusting for inflation and population growth in 1954, this 1939 figure is comparable to roughly USD 9,500,000,000 in 2019. According to the study, the ease with which homemakers may have moved some of their food preparation to the factory has resulted in a $4 billion rise in yearly spending since 1939. However, the real net cost of this tariff transfer may be larger. Cooked meals, as opposed to raw meals, significantly impact transportation and distribution costs. Weight, volume, and the type of storage required are only a few of the elements that influence these expenses. There are certain instances when they are less than others. Despite an increasing trend in food prices, food processing has consistently lowered expenses. Processing costs are most closely connected to raw material costs. To guarantee raw material supply, factories generally negotiate arrangements with local farmers. Farmers often accept lower agreed-upon prices rather than higher but fluctuating new market prices. Compared to the price of fresh asparagus in California in 1957, which was 272 dollars per ton, the cost of processing asparagus was 168 dollars per ton. Processing expenses can be kept to a minimum by using waste grades of raw materials. Produce with flaws is employed in production to save money, and this includes little oranges, malformed potatoes, massive sweet potatoes, deformed eggs, and apples with damaged skins. These flaws are insignificant after the raw material is peeled, sliced, juiced, or treated to other treatments. Raw materials that would normally spoil before reaching the client are stabilized and waste is reduced by processing. Waste and deterioration are normal byproducts of shipping and distributing fresh raw materials. The loss in the retail sector may be more visible, where much-overcooked food is frequently thrown away. The annual cost of this loss for fresh eggs alone is estimated to be $125,000,000. The elimination of waste materials during processing substantially decreases the costs of packing, storing, and transporting food from the farm to the market. The pod, which accounts for more than half of the total weight of a green pea or lima bean, is frequently left out of processed items. The weight of sliced sweet corn is much less than that of whole corn. During the milling process, 30% of the weight of the wheat kernel is lost. Fresh oranges weigh nearly twice as much as a can of orange juice. In some circumstances, such as canning, extra liquids such as sugar syrup or brine are added to replenish lost weight. Typically, a No. 2V2 can of peaches has 11 or 12 ounces of syrup and 18 ounces of peach halves. As a result, the total weight of the canned product given may be larger than the weight of the fresh peaches required to create the pack. (Chemat, F. 2021).

Costs can be minimized by fully using an item or animal. To maximize the value of an animal's life, most of its flesh is sold fresh, with only a tiny portion of the corpse processed into canned meat products and byproducts. Citrus peels may be used to make animal feed. Milk processing enables producing a wide range of valuable items, including butter, cheese, skim milk, milk powders, and whey while utilizing just a fraction of the original product in each case. Apples rejected by fresh-market packinghouses can be juiced, their peels used to create pectin, and the remaining cake fed to cattle. The value of raw materials may be raised by processing, and new products for specialized markets can be generated. The processor may improve the final output by selecting a high-quality component from the input material. Peach halves drenched in rich syrup may command a premium price. This nutrient is also found in white flour, chickpeas, and artichoke hearts.

When the expenses of maintaining a higher weight are revealed, efforts are made to lower them as much as possible. Concentrated products assist you in losing weight faster. One can of frozen orange juice concentrate produces the same volume as four ordinary orange juice containers. Before being frozen, dehydration decreases the weight and volume of fruits and vegetables by around half. Condensed canned soups can save up to half the weight and space. These enhancements significantly lower the cost of shipping, storage, and packing. Customers will save money if dehydrofrozen peas packaged for restaurants cost less than 2 cents per pound. There is a price difference between canned and frozen orange juice, albeit frozen concentrate is frequently less expensive. Even though it would be difficult to cover every technical innovation we could think of, we did.

Material handling and continuous processing improvements have resulted in considerable cost reductions. Water is a hydro-conveying fuel to carry goods through channels or open troughs. Hydro conveying is commonly used in the canning and freezing industries to transport perishable items such as fruits and vegetables. Green peas, lima beans, and sliced corn are just a few vegetables commonly transported through submerged pipes. Using high-tech water pumps raises the temperature of the combination to unprecedented heights. When most items are submerged in water for a lengthy period, the soluble solid components leach into the water, reducing their quality. Leaching losses are more prevalent in items with exposed surfaces, such as those that have been peeled or sliced. However, hydro transmission has several advantages that may outweigh its limitations. As a result, goods are relocated with the least amount of time and effort. As a result, the material is not allowed to dry out. It aids in chilling the substance to the degree where freezing is feasible. The more you order, the cheaper each item becomes. The material is transported in massive containers, vehicles, and tanks rather than small amounts. A bulk carrier, often known as a tanker, may transport dry and liquid goods. Many businesses use liquid sugar, carried in tank cars or trucks and pumped through a pumping system throughout the manufacturing process. Milk is often transported directly from farms in tank trucks and vehicles. Peanuts and soybeans are two bulk commodities besides cereals and peanuts. Produce is delivered in large quantities to processing plants. Totes with a ton-plus capacity are required for transporting raw materials to the plant. Large bins transport large dry goods such as flour throughout a processing plant. Lift vehicles are continually discovering new uses.

Totes or pallets of huge cases or boxes may be hoisted to heights of more than 20 feet, and this efficient equipment can transport them quickly across factories and loading yards. Pneumatic conveying, or the transportation of materials through air current, is increasingly being employed in processing powdered, crushed, and flaky commodities. Dry items such as flour, milk powder, sugar, beans, chocolate, coffee, peanuts, malt, and rice are moved using pneumatic conveyors. Objects are drawn into a conduit and exposed to high-velocity blowers. The duct might be hundreds of feet long depending on whether it's linking buildings, machinery, or storage rooms within one. In a collector, the product and the air are separated. While a result, the airflow is significantly decreased, and the product falls to the bottom as the air gradually rises to an exhaust vent. Because of these long-distance, uninterrupted transportation techniques, less money is spent on package processing. Traditional tactics need much physical effort to get the same outcomes as contemporary ones. As a result, the overall cost of the final product decreases.

In a large, well-organized processing facility, one worker may transport hundreds of tons of product. Factory food preparation is more cost-effective due to the high labor expenses connected with home cooking. Modern canneries frequently employ cutting-edge continuous cookers. There has been previous usage of the batch format. Constant production of dairy products such as butter, cheese, and ice cream is a viable option. Continuous diffusers increase sugar yield from beets while lowering sugar extraction labor costs. Baking equipment such as automatic bread deepeners and band slicers capable of cutting 50-60 loaves per minute is becoming more widespread. In the processing sector, a piece of machinery is deemed "continuous" if the inputs and outputs occur in an uninterrupted cycle with little to no human participation. A batch machine is analogous to a standard oven in the house. Each of these baked items is created by hand. In a continuous oven, the belt travels continuously down the length of the very long chamber. The burners are positioned and operated so baked goods may be constantly fed into one end of the chamber and retrieved as completed products. Citrus fruit juice may be extracted mechanically. Machines have mostly replaced humans when putting fruits like peaches, pears, and apricots into cans. To save processing time, exhaust boxes have been replaced with vacuum seals. Tomato paste, tomato juice concentrate, and other concentrates are frequently made in low-temperature, high-vacuum continuous evaporators.

The need for cutting-edge, always-on dehydrators is increasing. The belt trough drier makes drying fruits and vegetables simple and trustworthy. Increasing drying capacity does not need a large financial investment. Although many of these improvements resulted in considerable cost reductions in processing, they usually necessitated large investments in factory construction. The capacity of the facilities to handle large amounts of food effectively allows for low per-serving pricing. The cost of pre-packaged, processed food varies substantially. Now and again, the bundle price will match the per-meal pricing. Canned goods preserved in airtight containers may be rather expensive. The same may be said about liquids and their packaging. (Chemat, F. 2021).

The cost of a container can be decreased if an airtight seal is not required for keeping the products. Thus, paperboard boxes for frozen peas and cellophane bags for dried fruit are less expensive than tin containers. Innovative materials such as laminated foil and plastic film are almost airtight and cost the same as classic films and metal cans. Packing an aerosol can, or a bomb is expensive. Reduce the amount of water used in blending or adding to the meal to save money on packaging. Dehydrated mashed potatoes are around one-fifth the weight of raw and peeled equivalents. Orange concentrate accounts for around a quarter of the overall bulk of the juice. Portion control has recently gotten a lot of attention. A single package might include four servings of meat. In rare situations, a container of dry mashed potatoes may include two serving envelopes. Other examples are separate jars of jam and sachets of chocolate drink mix. While portion packing may raise container costs, it has the potential to reduce food waste in the kitchen.

Buying frozen, canned, or dry food offers both financial and environmental advantages. Whole-food dried foods are often the most cost-effective alternative for buyers. They're significantly smaller and lighter in weight. There is no need for low-cost storage temperatures. Packing specifications are usually unneeded. Dehydrated meals vary greatly in terms of cost-effectiveness. When vacuum drying is required, the price of dehydrated products rises, but output remains modest in comparison to the size and cost of the equipment. There are two specific instances when the lowest processed apples are frequently just dried apples. The price of vacuum-dried orange juice powder is likely to be higher than the price of the same volume of orange juice in a can or frozen meal. To prevent an increase in moisture content, the product must be hermetically sealed and dehydrated at a high vacuum, which is an expensive procedure. From a financial sense, there are advantages and disadvantages to purchasing canned vs. frozen food. (Chemat, F. 2021).

9.4 DISADVANTAGES OF FOOD PROCESSING TECHNOLOGIES

One disadvantage of the tincan is the increased price. Another disadvantage is the increased expense of transportation and storage of frozen food. The proportional relevance of these characteristics varies according to the nature of the product and the context. Because both canned and frozen peas are kept in tins, non-refrigerable condensed soups produced using canned peas will be less expensive than frozen equivalents of the same concentration. Whether or not the savings or increases in processing costs have been passed on to the client usually arises when discussing changes in processing costs. (Chemat, F. 2021)

Chapter

10 CHALLENGES TO BOTANICAL SCIENCES AND EXISTING OPPORTUNITIES

Contents

10.1 INTRODUCTION

The field of botanical sciences has been very useful in promoting the effective management of plants in various fields including agriculture, forestry and pharmacy. The various research conducted has provided very useful research that has enabled us as the world to know the position that we hold with regards to plant population. Despite the various and remarkable achievements, there are various challenges to botanical sciences all of which may be different depending on the area of applied botany. Some of the challenges are discussed as follows:

10.2 CHALLENGE IN PLANT OR TREE PIPELINE

Botanical sciences have led to the identification of various species of plants that were in existence before and have become extinct and those that are still in existence and are on the verge of going extinct. Under botanical forestry, studies show that in various countries across the world, the percentage of forest cover has reduced. This is attributed to the fast growth in human population which seeks new areas to settle. The depletion has also been attributed to the growing industrialization and the heavy industrial use of trees for timber among others. The use of timber as fuels in developing countries is despite the emphasis placed on the need to increase forest cover and various measures being put in place to bring back plant species that have already become extinct. One of the measures put in place was the introduction of a plant pipeline where various groups of people are involved in increasing plant forest cover. This measure though good, has not been highly effective and is greatly experiencing some draws backs hence the challenge on plant pipeline. (Stern, D. 2020).

This challenge is attributed to the fact that people do not have good views concerning plants. Most people view plants as the basis of life. This causes many to undervalue the importance and fascination of plants. This brings about the term plant blindness. Plant blindness is a real thing as it is demonstrated by the large number of people who do not put high emphasis on the planting of plants and trees. Plant blindness has been noted both in schools among primary and secondary students and also among adults. In third world countries, plant blindness is attributed to the fact that a significant number of the population have grown in rural areas and may not have gotten enough information on the importance of trees and planting them. Among school children, plant blindness could be attributed to the fact that it is being reinforced in school. It could also be attributed to the fact that the curiosity about plants and their importance has eroded. (Jha, P. (2017). In the early stages of botanical sciences, there was great concern for the environment. People were worried about climate change, the extinction of plant and animal species and the health of the environment. Therefore, the poor plant pipeline could be attributed to the fact that children are not being well taught and inspired to understand the importance of growing plants. The plant pipeline in the growing or crops is linked to the various stages of growth of a human being. Once a child is born, they go through early childhood education, primary education, secondary education, undergraduate, postgraduate and adulthood. If throughout the growth stages of the child, he or she is not taught about the importance of trees then there may be leakages in the plant pipeline. The leakages are noted in areas where there was no emphasis on the importance of plant growth. (Kopittke, P. M. 2016).

If children do not understand the fact that there are terms such as plant biodiversity and extinct plant species, then they may not show any concern in taking care of the environment. At the graduate or post-graduate level, the student has some knowledge of the importance of plant life. Some of them are studying the various biomedical sciences. They have the motivation to ensure the plant pipeline is well maintained. They can conduct various activities to ascertain this. Such activities include conducting tree planting activities or plant related seminars. This means that the greatest task is with children between primary to secondary education. (Stern, D. 2020).

There is a need for children to be exposed to the world of plant diversity. This may involve them learning various aspects including photosynthesis and the anatomy of plants. This has been boosted by the fact that some curriculums have already incorporated it into the education process. Various steps have been taken up to deal with such problems. One of the steps involved the holding of the symposium ' The Big Botany Challenge'. The symposium held in November 2019 was taken as the first step to dealing with the plant pipeline challenge. By nature, the symposium aimed at educating various groups of people on the different plant varieties. The symposium saw about, eighty presenters and delegates attending. They were from various conservation bodies, schools, research institutions, exam boards, and botanical gardens. Among the issues addressed in the symposium is how best plant profiles on the national curriculum can be improved in secondary schools. There were various suggestions including the adoption of the most suitable plant to be kept in the classroom and the development of career paths that will allow the diverse range of plant-related employment to become more accessible to both students and teachers. From the symposium held in the UK, their outcome was that there was a great need for the adoption of a strategy allowing the integration of a wider-ranging scope that will encompass the various parts of the educational pipeline (Jha, P. 2017).

From the symposium, it was highlighted that there was a plight of the British Botany. The plight was presented on the basis of an international level where it was noted that there was a kicking out of botany from the educational pipeline in Britain. It was noted that market forces took control over the botanical environment. Comparisons were made between China and the UK. China was noted to adopt an effective integrated approach under China's Strategy for Plant Conservation. The approach was noted to be very useful in promoting plant science research and education by making it a collaboration of disciplines. It was further noted that improved scientific capability within the society plays an important role in dealing with the problem of plant sustainability and species and ecosystem conversation. (Kopittke, P. M. 2016).

With regard to plant pipelines, some challenges were noted among women in science. The challenge was presented in the image of a leaky pipeline. Some of the noted challenges include the lack of encouragement or development of a positive perception towards Botany among schools hoping to start their career pathway in botany. By highlighting the various challenges, it became easy to identify some if the strategies that can be taken to effectively deal with the problem. For instance, the problem of lack of knowledge and interest by students in the classroom could be addressed by using plant stories. Children tend to be fascinated by stories. These plant stories can be used in the classroom to captivate the students. Plant stories can be used in other subjects including geography, English, chemistry and history. The illustration was with the knowledge of plant species found in the Amazon which can be pointed out through the use of various tools. They can be very useful in giving information on the Amazon and the kind of activities conducted there. The other solutions involved in the process include the use of wonderful plants such as the various fern relatives that children lay consider very interesting. When presented before children, they may develop the needed passion to enable them to get a better understanding of how the high levels of carbon in the atmosphere can be reduced. (Stern, D. 2020).

Another approach may be teaching botany by stealth. This can be achieved by telling plant stories of plants such as cotton. The stealth part comes about through the presence of shapes in plants. Plant interest among students can also be increased through the introduction of extraordinary plant structures including killer plants.

The other method involved allowing schools, students and teachers open up their career pathways. It was noted that there was a great need for the government and all the relevant stakeholders being involved to ensure that the needed awareness is created on the different types of plant-related career pathways. From the symposium, there was significant evidence to show that there was much to be done with regard to ensuring improved educational pathways to botany and plant health professions. Studies show that there are several benefits of better thinking through student exposure as they learn the subject. There was a great need for more research to be done with regard to motivation, science capital, raising awareness and generating various aspirations for science careers. Discussions from the symposium led to the conclusion that there was great need to develop and promote better links. This means that schools should have access to school career advisory services. This will enable them make better choices on what to study in higher institutions of learning. In most cases, these career paths are related to botany. (Jha, P. 2017).

There is also a great need for the ministry of education to make the best use of plants in the secondary curriculum. One of the things that contributed to leakages in the plant pipeline is the lack of emphasis on plants and their importance in the school curriculum. Some of the constraints were linked to the department of Education. This was noted by what was tested in the exam room. There was a limitation in exam boards not being able to maximize their opportunities such that they can greatly incorporate plant-related contexts. For instance, studies show that the use of context-led teaching can be useful in raising plant profile in biology. Among the resources that have to be utilized include the Science and Plant for Schools. It gives teachers the needed learning materials. The platform may include teaching students the role of plant science in citizenship science. One of the ways used in the evaluation of the existing plant-related content in secondary schools is the utilization of mind-mapping exercises. Such exercises have proven useful in helping students develop aspirations and students' science capital. (Stern, D. 2020).

From the symposium, it was noted that there was a need to adopt plant teaching resources and techniques in schools. They were anticipated to prove useful in developing interest among students. The same was noted with regards to the use of cool plants in very high temperatures classrooms.

These plants can be used as a teaching resource. The plants can be used as a good illustration tool. This is further backed up by the fact that Biodiversity learning is well understood when a sense of familiars is developed among students more so when dealing with botanical terms. There are various training resources that can be used by teachers. Teachers should also be allowed to develop gardens within schools. This enables students to get hands-on skills in botany among other things. Houseplants can also be used in the classroom as they are very important teaching resources. There are various plants that can be used while some species are preferred as they can survive in hot dry environments. They made the classroom look beautiful thereby introducing students to fun and beauty. Plants such as carrots cannot be grown as they may end up drying up within a short time.

The UK decided to hold the symposium more frequently so as to analyze the progress that has been made so far. In some countries, there may be a need to review the classroom curriculum so it fits the needs of the student. Some schools have made it compulsory for all students to learn biology. This means that every student is presented with an equal opportunity to interact with the field of Botany and gain more knowledge about plants. Practical lessons have also been adopted in some schools where the students are allowed to get samples of various parts of plants and study them under a microscope. This goes all the way in allowing them to get more information about the plant. (Stern, D. 2020).

10.3 LOSS OF PLANT BIODIVERSITY

Research from various organizations such as the Botanic Gardens Conservation International shows that loss of plant biodiversity is a major problem in botanical sciences. Their data shows that about 20% of plant species are under threat of extinction. It is anticipated that this will greatly affect human innovation, resilience and adaptation when dealing with various problems including climate change, biodiversity conservation, food insecurity, renewable energy, human health and water scarcity. High emphasis is placed on the need to understand botanical sciences such that research can be used to better manage and conserve the various plant species. The history of botanic garden sciences could be useful in dealing with problems related to the loss of plant biodiversity. This takes into consideration the legacy of the endeavors and the various challenges faced by botanic gardens. (Jha, P. 2017).

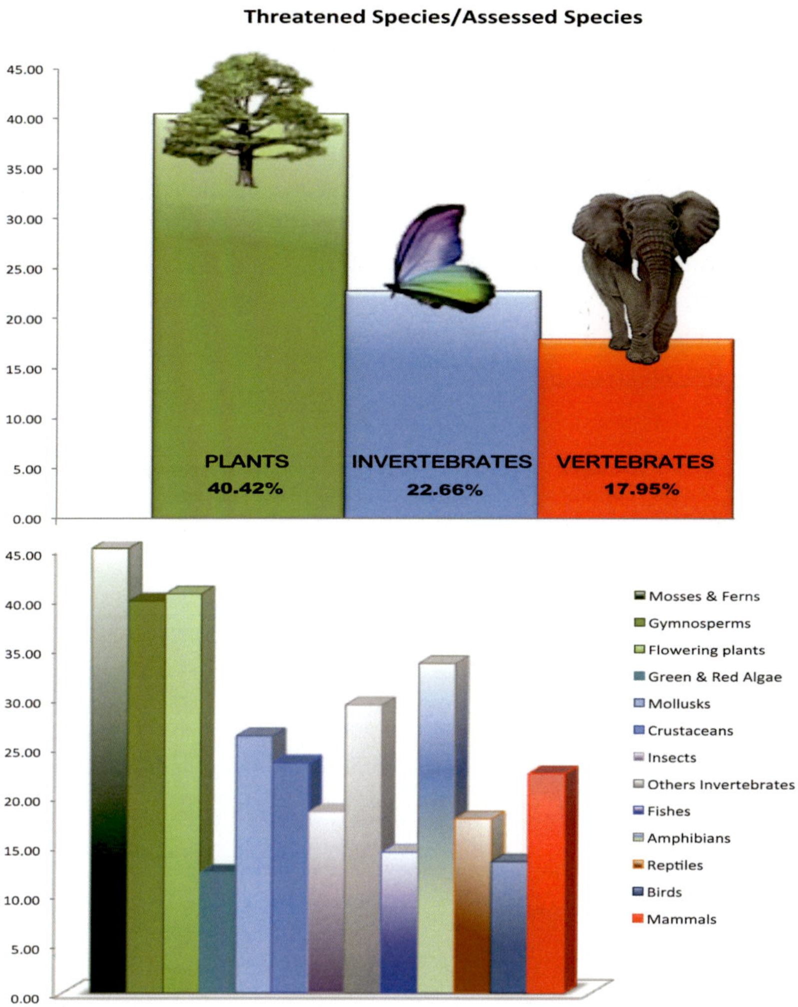

Figure 10.1: Loss of plant biodiversity is a problem in various parts of the world.

Source: Palombo MR. Thinking about the Biodiversity Loss in This Changing World. *Geosciences*. 2021; 11(9):370. https://doi.org/10.3390/geosciences11090370

In the current day and age, there has been a notable decline in plant biodiversity. This is indicated by the number of plant species that have now become exit. This has been noted more so among tree species that have been greatly lost through their industrial use. Exotic plant species mostly used for medicinal purposes have gone extinct as they have been greatly used in treating various diseases. There are various causes linked to the loss of biodiversity. They include changes in land use. Initially, most of the land was covered by a large percentage of tree coverage. But various activities such as urbanization will lead to a reduction in plant cover. This is because large tracts of land are cleared to allow the establishment of more houses that could allow humans to settle. Land is also cleared to allow the establishment of factories generating clothes and other tools. Land is also cleared through the process of intensive monoculture. This involves intensive farming of agricultural lands that leads to the depletion of minerals that should be useful in ensuring the survival of various plant species

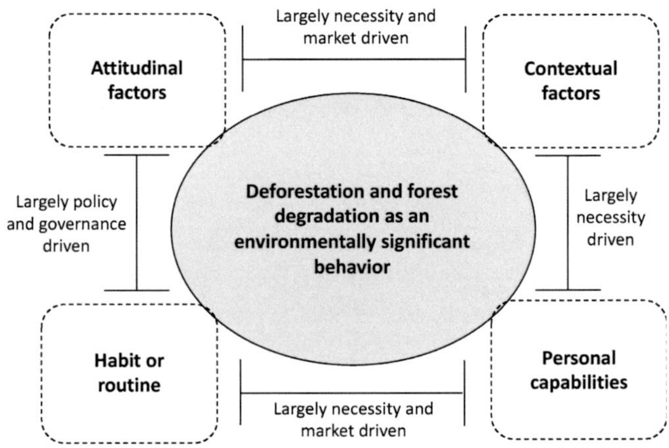

Figure 10.2: Deforestation has led to the destruction of various plant species.

Source: Duguma LA, Atela J, Minang PA, Ayana AN, Gizachew B, Nzyoka JM, Bernard F. Deforestation and Forest Degradation as an Environmental Behavior: Unpacking Realities Shaping Community Actions. *Land*. 2019; 8(2):26. https://doi.org/10.3390/land8020026

Deforestation also contributes to the loss of biodiversity in trees. Deforestation is defined as the excessive cutting of trees without replacing them. Some of the reasons why trees are cut are to provide wood for timber, use as fuel and as a source of raw materials. Deforestation has greatly affected exotic plant species. Most of these plant species cannot be reproduced as the most appropriate technology has not been developed. This means that most of these plant species head to extinction. Deforestation can also be placed in the same category as direct exploitation. This is because the plant species are directly extracted and heavily utilized. Direct exploitation can also encompass the utilization of plants as a source of food. With the fast growth of the human population, it is becoming more difficult to ensure a continuous supply of food. The heavy harvesting of plants has contributed to the loss of plant diversity more so in gardens as the lost plant species cannot be generated once more.

Climate change has also contributed to the loss of plant biodiversity. The frequent changes in climate have led to the drying up of lands such that most of the plant species cannot survive such harsh conditions. Harsh conditions such as strong winds and heavy rainfall may also cause the damaging of plant biodiversity. Such conditions do not allow the growth of such plants causing most of the plant species from heading to extinction. Pollution is another factor that has contributed to the loss of plant biodiversity. Certain plant species require very specific conditions to allow their growth. The presence of heavy contaminants in air, water or soil may undermine the growth of such plants. For instance, the presence of certain nutrients in soil in very large amounts may lead to the leaching of certain nutrients in plants causing their poor growth or eventual death of the plants. This greatly affects plant biodiversity. Another factor is the invasive alien species. They may take away the nutrients needed by plants. These species may inhibit the growth of various plant species. They may carry certain pests leading to the death of various species of plants (Jha, P. 2017).

There are various factors that contribute to the loss of plant diversity in various botanic gardens. Since time immemorial, the disciples of taxonomy and economic botany have dominated botanic garden science. The undertaking of research led to the generation of botanical gardens that would physically grow medicinal plants. This is the case with modern gardens. In the case of traditional gardens, the botanical gardens were driven by economic botany. Some of the gardens include Sydney, Bogor, Calcutta, Peradeniya and Kew among others. These botanic gardens were mostly involved with the growing and supply of cash crops including rubber, tea and coffee among other plant-based products. Most of these crops became popular due to the botanic gardens. The process of describing and collecting plant diversity in Botanic Gardens and was considered the means through which its end would lead to the development and production of new products. The continuous collecting, naming and describing of existing plant diversity eventually became the main scientific output. This means that there would be sufficient room to allow the discovery of new plants. The challenge however is that the process often at times would not transform the economy. This was noted in most university -based botanical gardens which developed in Europe. (Stern, D. 2020). Herbarium and living collections have over the years increased. They formed the basis of practice and teaching plant taxonomy. This led to many publications. In the current day and age, there are about 390,000 plant species that have been listed in the publications. This means that every year, about 2000 new species of plants were discovered. This allowed the development of a stable taxonomy for plants driven by exponential advances. This in biology is noted in information technology and molecular biology. This does not necessarily mean that there is no more plant taxonomy. It means that all the data are pointing towards the point where plant taxa are used in describing and the allocation of unambiguous names. It may be very necessary to look into the accumulated knowledge. Such information may be useful in dealing with the world's biggest challenges.

In the various measures put in place, it may be very necessary to note that the enormous body of taxonomic knowledge presented in the form of descriptions and names is very essential in ensuring the conservation, utilization and management of plant diversity. Initially, the existing relationship between the utilization and presence of taxonomy was not properly articulated. This is also a challenge in the current day and age. For botanists, taxonomy may not be pleasant because a lot of work goes into developing it. To some, taxonomy is very interesting as it gives significant information on the plant taking into consideration its properties, growing method, processing of plants and finally selling the plant product.

To develop the current day and age taxonomic grouping is linked to several dried herbarium specimens. These plant specimens have various roles including serving as a physical reference linking a name and description and a source of additional information on plants. Some vital information includes the occurrence of the plant, flowering and fruiting, uses and the various vernacular names. The taxonomic knowledge is also obtained from botanic gardens. Such gardens have been useful in building impressive living collections grown in the outdoors or in glass houses. Currently, botanic gardens have been said to produce seed banks of known plant species at 39%, 93% of vascular plant families and 59% of plant genera. Other additional data generated include wood collections, economic botany collections, DNA collections and carpological collections. They all play different roles in dealing with challenges that may be presented in the near future. Issues pertaining to the loss of plant biodiversity can be dealt with mostly by scientific institutions. This is because botanic gardens have the ability to answer the various questions pertaining plant biodiversity. This can also be attributed to the fact that for outsiders, the study and growth of obscure plants may not be of much importance to them. Some of them may consider it irrelevant and in some extreme cases an esoteric indulgence. However, there is a possibility of the indulgence of botanic gardens and the general population. The relationship was greatly linked to music events, light shows and aesthetic-orchid festivals. Though it may play a significant role in promoting interest, it has not met the needs of botanists in the means of promoting scientific research. Most of them may visit such gardens to marvel at their beauty but not to get more information about the plant (Jha, P. 2017).

This is also the case for all the governments and the relevant stakeholders who provide the needed resources to allow the establishment of various botanic gardens. Most of them tend to see such gardens as a place to visit and wonder at their beauty rather than the required view of seeing them as scientific institutions that play a vital role in dealing with problems of loss of plant Biodiversity among other environmental problems (Stern, D. 2020).

10.4 EXISTING OPPORTUNITIES TO PLANT BOTANIC SCIENCES

Plant botanic sciences are a broad field and only a fraction of it has been studied. There is still much to discover in the field of botany. One major opportunity that stands out is that it will be used to address major challenges affecting human beings. More information is being generated with the development of new technologies. For instance, at first, there was the normal ordinary microscope that allowed one to study the plant cell. More technological developments allowed the development of the electron microscope which allowed further studies of the plant allowing further distinctions between various plant species. This means that there is a great need for technological developments that will transform the world of plant biology and eventually allow the solving of world problems. Currently, plants work essentially to introduce energy into the biosphere. This means that it is essential for the production of food and other materials used by human beings. They are vital in shaping the environment. Studies show that three of the main problems affecting human beings in the current day and age are environmental degradation, energy and food. Plants are able to solve all these problems when the right strategy and approach are put in place.

Plants are central to food production. They contribute towards food safety either directly or indirectly. They have been well structured such that they allow the production of energy. They also influence climate change, habitat destruction, waterway pollution and agricultural expansion. All these problems are independent of each other. This means that there is a lot to deal with in the field of botanical sciences. For instance, any changes in climate put more stress on various habitats and food production. The good thing however is that more botanical research will allow these challenges to be solved currently and in future days to come.

Development of new technologies

When discussing the opportunities available in plant botanical sciences, technological development comes up as it allows advancements in plant sciences. Technological development is very useful in testing the existing hypothesis and the generation of new hypotheses and information. More advanced technologies are needed to enable a better understanding of the functioning of the essential organisms while ensuring that a sustainable future is built. These technologies are usually specialized such that they allow proper studying of cell structures. There are various technologies used in botanical sciences. They can all be placed into three main categories namely the currently existing technologies. Such kind of technology has already been developed and has been used for a number of years to generate the already existing information. The second category of technology is the new readily envisioned technology. This is what researchers have recently developed. The final category of technology is the technology that we would like to develop but have no means of development maybe because there isn't much information to allow the development of such technology. (Stern, D. 2020).

The various technologies are very useful as they make research possible. Some of the technologies already in use include electron microscopy, DNA sequencing, mass spectroscopy, RNA cataloging and fluorescence-based microscopy. The opportunities presented in botanical sciences are the development of new technologies more so technologies that can be used in imaging. There is a lot of hope with regard to the development of new technologies as various institutions have demonstrated interest in developing the needed technologies. However, it is anticipated that new technologies will eventually be developed and they will revolutionize the field of botanical sciences and research. For this reason, plant science is said to be a great driver of innovation.

Career options for students

The studying of plant botanical sciences is very beneficial to students. They can be able to work in various areas. For instance, some of them can work as plant biologists. There are various employers looking to hire plant botanists. Areas of employment include educational institutions, stage agencies and industries. Securing the job will be determined by the experience of that person and their educational training. With the continuous growth of botany as a field, it is expected that even more opportunities will come up. Introduction of new jobs is expected at an above-average rate. The growing population pressure will push for this. There is an increasing rate of food insecurity that pushes for the need to come up with more plant species that are resistant to various climatic conditions and to ensure a continuous food supply. Botany also looks into the environment by studying various aspects such as soil pollution, water and air. More research has to be done on these factors as they greatly influence plant botany. This means that there will be more openings for ecologists in the industries and in the government. There will be openings for those who are interested in studying the use of various plant species in treating various diseases. Genetic modification of crops is still an area that needs more research and hence another opening. (Jha, P. 2017)

It is also expected that there will be an increase in the number of learning institutions that will offer plant sciences. More students are looking into botany as a possible career path. The growing need for more research will push for the building and commissioning of new research institutions. More educational institutions are expected to employ a large number of plant botanists. With the need for knowledge of botany even at the elementary level, such botanists may also be hired by high schools or community colleges. They may be involved in the teaching of specialized courses while in those institutions though they may face the challenge of limited equipment. With more levels of education being introduced, more positions of professional plant scientists will be created. These scientists working in research institutions may work along with the government among other agencies to ensure that botany is well taken care of. They are able to work with various groups to fund their research. (Stern, D. 2020).

More room for research in plant science botanic garden

Plant Botanic Gardens are part of plant sciences. They are known for their great and longstanding botanic strengths in various areas including the exploration and documentation of existing and newly found plant diversity. This is achieved through the use of horticultural science, phylogenetics and systematics. Botanic Gardens have played a major role in providing most information concerning this topic. The challenge however is the fact that the development of mainstream plant science in the current day and age has led to the negligence kg botanic gardens. This may be attributed to the fact that the research process in mainstream plant science is mostly reductionist focused. This presents a challenge in the ranking of botanic gardens. Some of the activities undertaken in the botanic sciences include the ranking of garden plant science and a generous in general plant science within the community. Such evaluations usually involve gauging the significance of the work done by botanic gardens. This led to the conclusion that these gardens are very useful in conducting cutting-edge research and also dealing with various societal problems. With the change in society and various elements in society, there is a great need for botanic gardens. In the current day and age, global change is taking place at a very fast rate and has drawn the attention of world leaders who decided that there is a need for something to be done concerning the matter. This has led to the management of plant communities and sound stewardship becoming an international matter.

Some of the factors bringing about these changes include the burning and destruction of natural and wild plant life. Most of these plant species cannot be grown like other crops as they are freely given by nature to human beings. Such kinds of damages have contributed to a significant percentage of the amount of carbon released into the atmosphere. Such emissions are controlled by plants and trees but with the growing decrease in plant and tree coverage, it becomes more difficult. Botanical science through the gardening element is among the plant-based solutions facilitating sustainable living, improved human welfare and the repair and restoration of carbon sinks and lost plant biodiversity. This creates more emphasis on the need for such gardens. They offer hope at a time when chaos is all around and the ever-growing population makes the management of various ecosystems more difficult. (Stern, D. 2020).

This will lead to the establishment of more botanic gardens to offer more plant-based solutions to currently existing problems. The need for more such gardens is also championed by the principle of inter-generational equity. The principle requires that if there is a present environmental problem then the is need to deal with the problem. This principle is mostly applied in biodiversity. Challenges pertaining the loss of biodiversity should be dealt with beforehand so that the future generation does not deal with it. It also requires that more people be equipped with skills in plant biodiversity. These people will be required to come up with innovative ways of dealing with problems of plant biodiversity. Usually, the solutions offered tend to be inspirational and innovative in nature. The new cadres may have to be stationed in certain places and chosen according to the needs of the region and of the people. Usually, a significant number of people are posted in a given area to enable effective research and production of food. The nature of work will not allow the one-size-fits-all kind of technology to be used to fix the existing problems. Such kind of technologies may not be effective in facilitating a sustainable future. (Jha, P. 2017)

Despite the various challenges such as those noted in the environment, it is worth noting that there has been an improvement and flourishing of biotic plant gardens. More people are becoming more interested in biodiversity collections, the form of research and the manner of utilization. Since 1950, several Botanic Gardens have been established. About half of the 2500 well-known botanic gardens were created in the early stages. Such a revolution has been linked to a strong sense of cultural relevance and existing technologies. Both elements are still useful in the current day and age in ensuring that there is a continual international passion to allow the creation of more botanic gardens. The development of new technologies will also allow the creation of new botanic gardens with the required plant species with the required characteristics. The newly developed gardens exhibit a sense of imagination and scale. At the very core of botanic gardens are the extraordinary creative designs and public outreach. Some of the recently developed botanic gardens include the Australian garden of the Royal Botanic garden, The Eden Project in Cornwall and botanic gardens in Singapore. The botanic garden in Singapore has not been completed as it is anticipated that an excess of 1 billion dollars will be used in the garden. The garden has been developed using government funds. Such an amount is really hefty for building a garden. For this reason, there may be a form of resistance or giving up when building such gardens causing people to resort to the traditional methods and services provided by botanic gardens. The increased cost may be due to the newly discovered information, investigation, collection, celebration and conservation of plant biodiversity.

It was also noted that more interest was developed in the construction of botanic gardens. Some years are very significant with regard to botanic gardens as the most venerable botanic gardens were established. The Royal Botanic Garden Kew is very old such that it has already celebrated its 250th anniversary while both the Singapore and Missouri Botanic Gardens celebrate their 150th anniversary. The success of these gardens was linked to the strong sense of community support, local, national and international support. The start of this also coincidentally occurred with the publication of the birth of Charles Darwin. This piece of information is very important as it allowed a proper explanation of how plant diversity came about. It has allowed the safeguarding of plant life for future generations. It has allowed the development of thousands of plant species. (Stern, D. 2020).

Scientists in the world today also make use of the information given by Charles Darwin. Scientific information has formed the basis of research for many plant species. This means that scientific information is really important. The development of the new types of gardens makes use of the collaborative and authoritative sharing of knowledge. Skills and a great sense of knowledge are also needed alongside the various networks in the various botanic gardens. The information generated by one botanic garden is also used by another botanic garden. More programs are also being developed. Most of them aim at creating more awareness of botanic gardens. Some of the programs are championed but the science-based organization.

More policies are also being made with regard to conservation due to the presence of botanic gardens. Despite this, there is a challenge with regard to the translation of international policies and commitments into well-required outputs. The process has been noted to take place at a very slow rate. Some of these commitments include the Global Strategy for plant conservation. This may negatively affect those involved in the process. Since the establishment of the various Botanic Gardens, there has been a notable increase in the collection of information for conservation and generating a good sense of understanding. This means that there is the hope of the possibility of leaning towards digital capture of the generated information. The needed steps are also underway but facing some setbacks with regard to the quantity of primary data which has become a bit difficult to resource.

To further allow the growth of botanic gardens, several studies have been done. Most of them demonstrate the potential and power of the highlighted approaches. This means that large amounts of data that have been generated will have to be consolidated to allow proper conservation and sustainable living. Currently, the new focus of botanic gardens is to allow the expansion of the traditional focus of botanic gardens through the patterns of diversity at the basic level (species. This allows the incorporation of the genetic approach allowing the delivering of real advances to allow the restoration and conservation of genetics and ecology. (Stern, D. 2020).

There is still more work to be done to allow adaptive genetic diversity. This will allow the strengthening of existing strengths in the areas of documenting variations in certain plant markers. Studies show that when the various plant species are subjected to changes in the environment, they may be able to survive the phenotypic plasticity while others may head towards extinction. The neutral markers involved in the study are not sufficient to allow the generation of information on challenges in conservation. This means that as more technologies are being developed, botanic gardens need to adopt ecological genomics. Such a kind of an adaptation will allow the use of important infrastructure to test different hypotheses demonstrating the various responses of various plants to the various changes.

If no drastic measures are taken, it is expected that the current common will become the new rare in the future. This is due to the fact that significant changes will affect ecological relationships. There is also a need to understand that the one size fits all approach may not apply when trying to get a better understanding of response to plants in various dynamic circumstances. This makes restoration ecology the new focus in botanic gardens. It is linked to other issues. The challenges faced in botanic gardens should not be underestimated. Even the least important part of plant species could spring out to be very important. This was noted by the vulnerable and biologically specialized taxa present hope for restoration only if the needed resources and innovation are applied in plant Science. There are various critical areas in botanic gardens and plant Science among them being seed banking. The challenge however is that it often at times is overlooked. Seed banking is very useful in giving more information and advancements needed to allow a proper understanding of seed ecology and biology. More discoveries are being made including the presence of certain chemicals in smoke that will boost the germination of various plants.

Different measures are applied depending on the nature of the situation. For instance, biotechnological approaches are used in tropical regions where plant species have recalcitrant seeds. This means that botanical gardens in tropical regions need to place more emphasis on making use of situ conservation and restoration programs. Botanic gardens also make use of citizen science. Citizen sciences have contributed to the large number of people visiting those areas with online or physical visitation. In the current day and age, so many developments have been made such that there is no more room for complacency. Botanists need to invest more in capitalizing and using the very unique opportunities that will enable them to better communicate with visitors and also inspire them. Botanic garden cycles bear similarities to other organizations as they experience periods of retrenchment and rapid development. (Stern, D. 2020).

It is important to note that there is more to botanic gardens more so in terms of plant-based methods of problem solving. They have become more relevant with dealing with different human problems. The government and other stakeholders need to develop a strong sense kg communication and advocacy more so with regard to border society. All these efforts need to be accompanied by unstoppable support of training, sustainable use, flourishing plant diversity science and conservation. There is still a great need for more optimism in the field of botanic gardens more so in plant Science. This will bring about more success in various ways. Botanic Gardens have also been involved in the development of different plant species that can be used in treating different medical problems. Botanical studies are being done to identify such kinds of plant species. (Stern, D. 2020).

REFERENCES

1. Adams, C. R. (2012). *Principles of horticulture*.London. Routledge.
2. Baldassarri, F. (2017). Introduction: Gardens as Laboratories. A History of Botanical Sciences. *Journal of Early Modern Studies*, *6*(1), 9-19.
3. Bendre, A. (2010). *Practical botany*. Deep and Deep Publications.
4. Bernt, M., Braband, A., Schierwater, B., & Stadler, P. F. (2013). Genetic aspects of mitochondrial genome evolution. *Molecular phylogenetics and evolution*, *69*(2), 328-338.
5. Borer, H. (2014). The category of roots. In *The Roots of Syntax*. Oxford University Press.
6. Bunma, S., & Balslev, H. (2019). A review of the economic botany of Sesbania (Leguminosae). *The Botanical Review*, *85*(3), 185-251.
7. Chandra, S., Lata, H., & ElSohly, M. A. (Eds.). (2017). *Cannabis sativa L.-botany and biotechnology*. Springer.
8. Charlesworth, B., & Charlesworth, D. (2010). Elements of evolutionary genetics.
9. Cochet, H. (2015). *Comparative agriculture* (pp. 154-p). Dordrecht: Springer.
10. Cook, A. (2017). Jean-Jacques Rousseau and botany.
11. Dávila, P., Soto-Trejo, F., Rodríguez-Arévalo, I., Ponce, A., Arias, S.,

Escalante, A., ... & Lira, R. (2022). Wild plant conservation in Mexico in the 21st century. *Botanical Sciences*, *100*(SPE), 175-197.

12. Eisenman, T. S., Churkina, G., Jariwala, S. P., Kumar, P., Lovasi, G. S., Pataki, D. E., ... & Whitlow, T. H. (2019). Urban trees, air quality, and asthma: An interdisciplinary review. *Landscape and urban planning*, *187*, 47-59.
13. Fagnani, M. L. (2022). From 'pure botany' to 'economic botany'– changing ideas by exchanging plants: Spain and Italy in the late eighteenth and the early nineteenth century. *History of European Ideas*, *48*(4), 402-420.
14. Gonsalves, J. (2010). *Economic botany and ethnobotany*. Mittal Publications.
15. Grace, O. M. (2011). Current perspectives on the economic botany of the genus Aloe L.(Xanthorrhoeaceae). *South African Journal of Botany*, *77*(4), 980-987.
16. Hall, M. (2011). *Plants as persons: A philosophical botany*. Suny Press.
17. Hardy, G., & Totelin, L. (2015). *Ancient botany*. Routledge.
18. Hardy, G., & Totelin, L. (2015). *Ancient botany*. Routledge.
19. Hilali, M., El Hammari, L., El Monfalouti, H., Maata, N., & Kartah, B. (2022). Proceedings On Applied Botany, Genetics And Breeding. *Proceedings On Applied Botany, Genetics And Breeding*
20. Insel, P. M. (2014). *Nutrition*. Jones & Bartlett Publishers.
21. Jain, S. K. (2010). *Manual of ethnobotany*. Scientific publishers.
22. Kartal, C. (2016). Calcium oxalate crystals in some species of the tribe Cardueae (Asteraceae). *Botanical Sciences*, *94*(1), 107-119.
23. Kochhar, Suraj Lal. *Economic botany*. Cambridge University Press, 2016.
24. Kumar, J., Choudhary, A. K., Solanki, R. K., & Pratap, A. (2011). Towards marker-assisted selection in pulses: a review. *Plant Breeding*, *130*(3), 297-313.
25. Liakos, K. G., Busato, P., Moshou, D., Pearson, S., & Bochtis, D. (2018). Machine learning in agriculture: A review. *Sensors*, *18*(8), 2674.
26. Lin, L., Ni, B., Lin, H., Zhang, M., Li, X., Yin, X., ... & Ni, J. (2015). Traditional usages, botany, phytochemistry, pharmacology and toxicology of Polygonum multiflorum Thunb.: a review. *Journal of Ethnopharmacology*, *159*, 158-183.

27. Lindley, J. (2015). *An introduction to the natural system of botany*. Cambridge University Press.
28. Magallón, S. (2014). A review of the effect of relaxed clock method, long branches, genes, and calibrations in the estimation of angiosperm age. *Botanical Sciences*, *92*(1), 1-22.
29. Magdoff, F. (2012). Food as a commodity. *Monthly Review*, *63*(8), 15-22.
30. Martin, G. J. (2010). *Ethnobotany: a methods manual*. Routledge.
31. Mauseth, J. D. (2014). *Botany: an introduction to plant biology*. Jones & Bartlett Publishers.
32. Mead, R., Curnow, R. N., & Hasted, A. M. (2017). *Statistical methods in agriculture and experimental biology*. Chapman and Hall/CRC.
33. Misra, N. N., Koubaa, M., Roohinejad, S., Juliano, P., Alpas, H., Inácio, R. S., ... & Barba, F. J. (2017). Landmarks in the historical development of twenty first century food processing technologies. *Food Research International*, *97*, 318-339.
34. Naik, K., Mishra, S., Srichandan, H., Singh, P. K., & Choudhary, A. (2020). Microbial formulation and growth of cereals, pulses, oilseeds and vegetable crops. *Sustainable Environment Research*, *30*(1), 1-18.
35. Pederi, Y. A., & Cheporniuk, H. S. (2015, October). Unmanned aerial vehicles and new technological methods of monitoring and crop protection in precision agriculture. In *2015 IEEE International Conference Actual Problems of Unmanned Aerial Vehicles Developments (APUAVD)* (pp. 298-301). IEEE.
36. Ren, H., Koshy, P., Chen, W. F., Qi, S., & Sorrell, C. C. (2017). Photocatalytic materials and technologies for air purification. *Journal of Hazardous Materials*, *325*, 340-366.
37. Režek Jambrak, A., Nutrizio, M., Djekić, I., Pleslić, S., & Chemat, F. (2021). Internet of nonthermal food processing technologies (IoNTP): Food industry 4.0 and sustainability. *Applied Sciences*, *11*(2), 686.
38. Shipunov, A. (2018). Introduction to botany. *Lecture notes. February*, *8*, 2018.
39. Slavin, J. L., & Lloyd, B. (2012). Health benefits of fruits and vegetables. *Advances in nutrition*, *3*(4), 506-516.
40. Song, R. B., Wu, Y., Lin, Z. Q., Xie, J., Tan, C. H., Loo, J. S. C., ... & Zhang, Q. (2017). Living and conducting: coating individual bacterial cells with in situ formed polypyrrole. *Angewandte Chemie*, *129*(35), 10652-10656.

41. Sparks, Donald L. *Advances in agronomy*. Academic Press, 2012.
42. Szafer, W. (Ed.). (2013). *The Vegetation of Poland: International Series of Monographs in Pure and Applied Biology: Botany* (Vol. 9). Elsevier.
43. Van der Putten, W. H., Bardgett, R. D., Bever, J. D., Bezemer, T. M., Casper, B. B., Fukami, T., ... & Wardle, D. A. (2013). Plant–soil feedbacks: the past, the present and future challenges. *Journal of Ecology*, *101*(2), 265-276.
44. Van der Veen, M. (2014). The materiality of plants: plant–people entanglements. *World archaeology*, *46*(5), 799-812.
45. Van Klompenburg, T., Kassahun, A., & Catal, C. (2020). Crop yield prediction using machine learning: A systematic literature review. *Computers and Electronics in Agriculture*, *177*, 105709.
46. Wang, P., Lombi, E., Zhao, F. J., & Kopittke, P. M. (2016). Nanotechnology: a new opportunity in plant sciences. *Trends in plant science*, *21*(8), 699-712.
47. Wickens, G. E. (2012). *Economic botany: principles and practices*. Springer Science & Business Media.
48. Wisecaver, J. H., & Hackett, J. D. (2011). Dinoflagellate genome evolution. *Annual review of microbiology*, *65*, 369-387.
49. Zadoks, J. C. (2013). *Crop Protection in Medieval Agriculture: Studies in pre-modern organic agriculture*. Sidestone Press.
50. Zörb, C., Geilfus, C. M., & Dietz, K. J. (2019). Salinity and crop yield. *Plant biology*, *21*, 31-38.

INDEX

C

D

H

I

L

M

N

R

S

T

U

V

W

Z

Kolibri 360 GmbH
Ferdinand-Jühlke-Straße 7
99095 Erfurt, Deutschland
produktsicherheit@kolibri360.de